Practical
Academic

MANAGING RESEARCH GROUPS AND PROJECTS

Effective management practices

for academic researchers

Jennifer Rowland

For information contact Jennifer Rowland

http://www.practicalacademic.com

Book cover design by Bek Pickard from Zebedee Design.

ISBN-10:0-646-96725-8

ISBN-13:978-0-646-96725-7

First Edition: February 2017

10 9 8 7 6 5 4 3 2 1

Foreword

Thank you so much for buying this book!

I developed this content over a long period of time spent reflecting on my experiences in academic research, as well as those of my peers, spanning multiple countries and institutions. Since stepping into a career focused on training postgraduate researchers, I have observed that many of the issues that junior researchers confront, could be better addressed at the supervisory level. In particular, if group leaders were to run their research more systematically, and provide a solid framework within which their team could operate, so many of the issues that I confront with trainee researchers down the line could be avoided.

The purpose of this guide is to give you a hands-on outline of what is it like managing an academic experimental research group, and some tips on everyday issues that you can encounter in your work. This book is mostly a practical guide, focused on helping you to identify managerial approaches that you can routinely implement in your role as a group leader.

The sixteen chapters of this book are divided into general group leadership (chapters 1-8) and project management (chapters 9-16). I have included **case studies** throughout, to illustrate concepts and provide insight into situations that arise in the research environment. **Exercises and reflection points** are included to the end of each chapter to relate the introduced concepts to your own experiences. **Downloadable materials** are listed at the end of each chapter: these can accessed from my companion Practical Academic website (www.practicalacademic.com/resources). These materials collectively form a "Group Leader's Toolkit", which can help you to simply employ the practices outlined here in your own group.

I sincerely hope that the materials presented here can help you to reflect and improve on your own academic research leadership practices. Many of you will naturally employ many of these techniques already, and thus may find this common sense, but the intention here is to help those that do not already consider these approaches. I welcome you to this first edition, and look forward to reading your feedback, as much like a quality Group Leader, I am always looking for consistent and ongoing improvement.

With warm regards,
Jennifer Rowland

Table of Contents

Disclaimer

The material and opinions presented in this book are my own and not representative of any other party or organization. While all attempts have been made to verify the information included to this book at the time of writing, the author assumes no responsibility for errors, omissions, or contrary interpretations of the subject matter herein. The author therefore disclaims any liability to any party for any loss, damage, or disruption caused by errors or omissions, whether such errors or omissions result from negligence, accident, or other cause. Adherence to all applicable laws and regulations, and all other aspects of doing business under any jurisdiction is the sole responsibility of the purchaser or the reader. The author assumes no responsibility or liability whatsoever on behalf of the purchaser or reader of these materials.

Acknowledgements

This book would not be a reality without the ongoing encouragement and support of Guy and Michaela Windsor, to whom this book is dedicated. So many others have supported its creation along the way. For sharing their critical eye, I am immensely grateful to Kaisa Silander, Josien de Bie, Sarah McNicol, and Fiona MacDonald. Huge thanks to Becca Judd who has provided outstanding proofreading of the manuscript, and Bek Pickard who designed the cover. John Rowland and Tracey Mackney provided superb general feedback, thankyou. Thank you to my friends who provided feedback on the aesthetics: Laura Quinton, Sara Happanen, Sam Clayton, Terri Campbell, Yvonne Barrett, Melanie Stanton, Gerry Stewart, Alicia Coyle, and Sally Purcell. Margaret Rowland, your unwavering encouragement has spurred the creation of this title forward, my deepest gratitude. My beautiful children Rafael and Maya have been incredibly understanding and supportive during the creation of this book, for which I am immensely grateful. To my friends and family, thank you so much for your incredible support, it means the world to me.

SECTION 1: MANAGING A RESEARCH GROUP

In this section we focus on overall group management, which represents the basis for establishing a functional research platform. We consider the development of a team and operational approaches that support a solid and innovative research environment. We also review common issues that surround the management of research groups. This discussion is presented in the following chapter structure:

- **Chapter 1** introduces the research business and how you fit in it as a group leader.
- **Chapter 2** describes several different group management structures.
- **Chapter 3** discusses issues relating to securing and retaining exceptional staff and students.
- **Chapter 4** reflects on practical organizational issues relevant to group management.
- **Chapter 5** discusses networking, both inside and outside of the institution.
- **Chapter 6** describes scheduling approaches and concerns that impact group management.
- **Chapter 7** focuses on common issues regarding personal and professional dynamics.
- **Chapter 8** introduces some approaches to help you juggle teaching commitments.

This section is delivered from the point of view of a group leader who is working in an academic research institution or university. The principal type of research discussed throughout this book is assumed to be experimentally-focused, predominantly quantitative, and pursuing hypothesis-testing investigations. Nonetheless, those pursuing alternative investigation styles may still benefit from the concepts discussed here.

CHAPTER 1

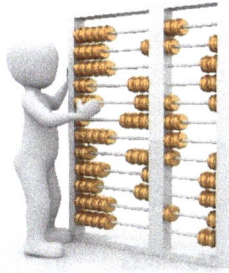

Running a Research Business

Research is a business, and considering it this way is advisable when you are part of a research group, regardless of your position within the framework. Nonetheless, as a researcher you are unlikely to be focused on revenue accumulation or profit margins, but rather will be aiming for particular project goals that might include journal articles, theses, or the development of something useful for humanity.

In order to achieve all of these awesome goals, you need to be organized and employ a range of tools. Although this is not evidently clear when you start training in research, it is definitely something that you are better off being aware of in the early stages of your work. You have to have strategic alignment. Financials are important. The people that you will be reporting to within your funding body want outcomes that they can be proud of, and they want effective use of funds within your project.

Overall, you have to **manage** your projects like a business, **report** on progress, **market** your research, **communicate** your research, and actually **perform** your research. You need to **liaise** with other participants. You need to **manage** your group, **support** your staff and students in their work and professional development, and **promote** a positive working environment to maximize the welfare and productivity of your group. If you are a part of an academic institution, you also need to deliver **teaching and mentoring** to tertiary students.

Remember, everything else is null and void if you do not achieve your research goals. You are required to wear many hats in this endeavor (figure 1.1). This book is intended to give you a blueprint for issues that you should remember to address in the establishment and management of your group, and in projects that constitute your research business.

Figure 1.1: The many roles of a group leader in research

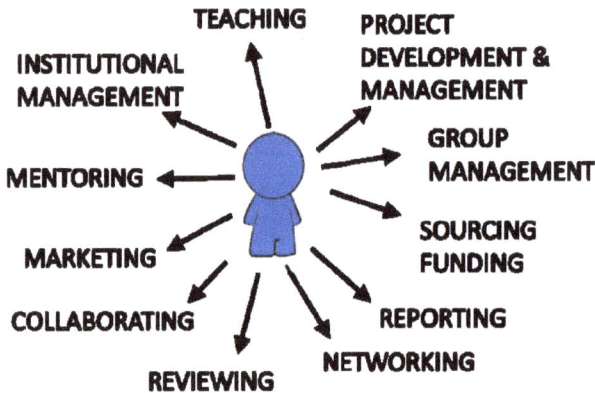

The Academic Business

Institutional recruitments have evolved in recent years to focus predominantly on the **"value added"** that a group leader might provide. This typically takes the form of how much funding equity a group leader brings with them, but it can also relate to a highly-desired skill or service that the department wants to acquire, such as a certain technique or facility. Whilst academic pursuits traditionally reflect advancement of knowledge, the focus is increasingly about trending technology, trending investigations, and — most importantly — the money your research can attract. A well-managed group that delivers consistent and high-level funding awards, in addition to high-impact research outputs, can excel in this environment. The flip side is that many highly original, creative, and innovative researchers simply grow tired of the relatively low pay versus the amount of dedication and time commitment required to meet all of their professional expectations in this rapidly evolving sector.

The only way you can ensure that you will succeed in this environment is to set up a successful and funded research group, in addition to selling the value of your work to the relevant institutions and funding bodies.

WHY DOES YOUR INSTITUTION VALUE YOU?

The Research Group Business Model

A research business model can be adapted from simple economic business structures that are well established in the competitive market economy. Figure 1.2 depicts a representation of a simple input/output transformation business model (Betz, 2002). It is always good to remember that, regardless of the circumstances surrounding the work that is being performed in any group, the overall task of a research group is to develop a deliverable by transforming resources; just like a business transforms resources into marketable goods, services, and knowledge.

Figure 1.2: Standard simple business model

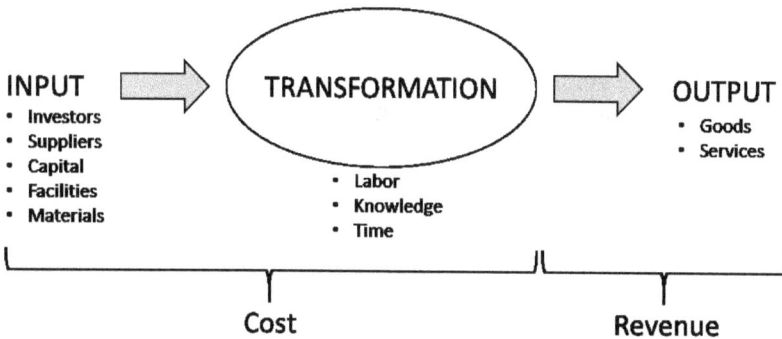

Of course research groups are very dynamic, receiving a broad array of inputs and developing a huge variety of outputs. They can serve a range of complex purposes, which are perhaps better represented by figure 1.3.

Figure 1.3: Simple research business model

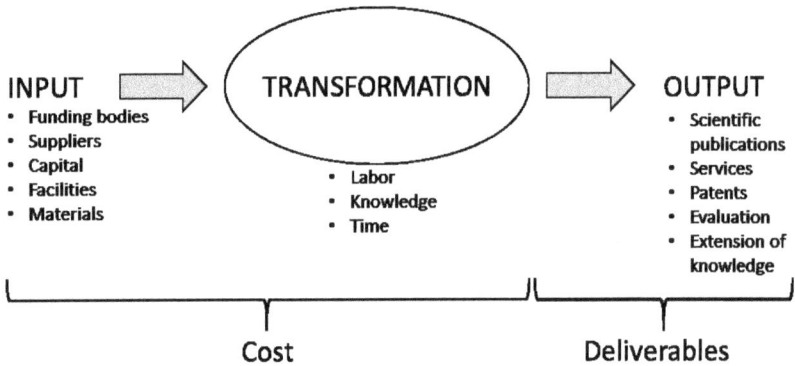

INPUT
- Funding bodies
- Suppliers
- Capital
- Facilities
- Materials

TRANSFORMATION
- Labor
- Knowledge
- Time

OUTPUT
- Scientific publications
- Services
- Patents
- Evaluation
- Extension of knowledge

Cost

Deliverables

The difference in the research business model (i.e. figure 1.3) is mostly derived from the expected output, which is not usually required to generate direct revenue — although this is not a rule. Researchers may also provide a service in tandem with their core innovative research, which can represent the value that the group brings to its institution.

Input and output in research

Groups can be funded by a range of sources, including standard government grants, private beneficiaries, charities, organizations, individuals, industry, host institutions, or scholarship funds. This list is by no means exhaustive.

There are a variety of outputs that may emanate from an academic research group, but the predominant focus of most groups is to expand the current state of the art relative to the focal research area for the group in question. This is a limited view, however, given the expansion of the academic sector over recent years to reflect a more business-focused enterprise. Groups are becoming increasingly diversified in the materials that they develop, with more emphasis on developing tools and specific services as contractors to other businesses or organizations. The development of patentable and potentially marketable technologies is also becoming a major focus for many academic researchers.

Resources

The means by which the group transforms the resources at their disposal into some feasible output is critical, if the results are considered to be reliable and true. Most experimental research groups generate quantitative data that requires solid statistical analysis, and their work should be able to be replicated by others.

The resources available within an institution, or within an individual research laboratory or establishment, are often critical to the success of the project that you have underway, particularly in quantitative experimental-focused research. Therefore, when you are establishing your research business it is critical to ensure that you are working within an environment that can deliver all of the required resources for your investigations to take place. Similarly, if your work is branching into a new direction and you require new resources that cannot be sourced in your current location, you need to expand your business. You can do this either by relocating (moving to a new institution), outsourcing (paying someone to do this specific new work on your behalf), or collaborating (networking with other institutions to achieve the work off-site or to access necessary resources).

Processes — labor and knowledge

The processes that any group undertakes to transform their inputs to deliverable outputs will vary as much as research foci vary, and every group will be different. Nonetheless, the ways that you manage your group, your research investigations, and your time are fairly similar. Different approaches to managing your processes are discussed throughout this book.

Defined goals and deliverables

Are you hoping your publications will lead to funding, promotion, growth, or the expansion of your research profile? It is a good exercise to consider how you would define your overall research goals in the context of this model. Are you willing to sacrifice time spent on pure research in order to secure the funding to keep your group afloat financially? Do you have a broad array of goals that you wish to achieve, which are dependent on individual project goals? Do you compartmentalize your work into separate task foci? Defining the expected outcomes for your group is key to systematically pushing the work forward.

Map Out Your Own Research Business

Creating a spreadsheet or document outlining your current research business can help you to define how it is structured. Here is an example created for a fictitious group (table 1.1).

Table 1.1: Mapping your research business

DATES	INPUT	GOAL	PROCESS	YIELD	STAFF
Jan-16 to Dec-16	Departmental Start-up Fund	Lab set-up	Organization of laboratory	Facility set-up	ALL
Jan-16 to Dec-17	Red Cross Innovation Grant	Potential Human Blood Therapy	Evaluating human blood-derived stem cells	Paper	Postdoc (LS) Tech (TN)
Jun-16 to Jun-18	International Fellowship IIF	Two-year project	Analysis of mouse model of dystrophy	Paper	Postdoc (NM)
Jan-16 to ongoing	Operation Funds	Provide TG Services	Technician - microinjection service for institute.	Service provision	Tech (AT)
Jun-16 to Jun-19	PhD Scholarship - Lab Costs	Doctoral Thesis	Signaling pathways *in vitro*	Paper	PhD Student (FS)

12

Finances

The above table is a great outline for systematically developing a business plan for *anyone* working on a project, and in any business venture. This type of outline is useful to include in future funding applications or institutional reports to demonstrate your group is afloat financially, and to show that you manage your varied interests effectively. It is advisable to keep an up-to-date outline similar to that above as a resource at your disposal. You may also include extra column headings that address your institutional account code for that fund, or to list who is responsible for managing the fund. How the respective balance sheets are managed is specific to the institution in which you are situated, and how they have chosen to manage their financial system. You should encourage your group members to be involved in the management of their own budgets, so that you enable them to become professionally-aware researchers who do not overspend your budget. Who knows when you might encounter a personal issue that takes you away from your work for a time; knowing that your people are able to manage the budgets in your absence is reassuring, given you have already built trust over time. This is covered more in chapter 4.

Institutional Services

In any role that you are associated with in a research group, it is important to understand the institutional structure within which you are operating. Understanding the structure, and most importantly the people who provide different services available to your research group, is critical to operating effectively within your environment.

When welcoming new staff or students to your group, it is worth sharing a **database of resources** for reference as they familiarize themselves with the new environment. This should be retained in a central information repository (paper or digital) for all group members and updated regularly. Some things that you may include are listed here.

Potential inclusions to database of resources

- Computing Services
- Outreach and communication
- Contracts and HR
- Departmental services
- Secretary
- Administrators
- Core facilities
- Postgraduate conveners
- Counseling services

- Accounting
- Learning and teaching
- Available equipment
- Careers services
- Counseling services
- Staff and postgraduate clubs
- Office of research integrity
- Postgraduate studies office
- Statistical support
- Printing services

In compiling your group's database, you should make a spreadsheet where you note the relevant contact, services provided (if not obvious), relevant web links, user names, and any passwords for your group. An example of a summary of available resources is shown in table 1.2 at the end of this section, which forms the template for exercise 1.2 at the end of the chapter.

Being part of an institution can pass on benefits like discounts from certain suppliers; regular networking opportunities, both social and professional; access to sporting facilities (to keep your mind and body fresh); transport discounts; journal subscriptions; and other library-related services. Institutions can provide in-house funding opportunities and established links with other institutions. Students can also often bring extra benefits through discounted services, such as document editing and revision, purchasing discounts, and user discounts via free student memberships. These are all things to take into consideration when you look at taking up a new post at any level as an academic researcher.

Student-specific institutional resources

Having students in your group who are enrolled in various academic institutions might provide extra resources that you were unaware of previously. These include access to the library catalog at the institution in which the student is enrolled, access to subsidized language revision services for manuscripts on which the student is included, or scholarships and awards specific to students who are attending their given educational institutions. Some institutions also provide a study space that the student can secure for the final stages of their research studies, in order that they may write up their thesis. It is worth asking your new students to research what they are entitled to, especially if you are hosting a student from a previously-unaffiliated institution.

STUDENTS CAN ADD VALUE TO YOUR RESEARCH BUSINESS

Facilities

If you consider your research group to be a business, you have to consider your research facility and offices to be your premises that you have secured by your affiliation with your institute. You may think that the allocation of your working space and office space is dependent on the institution with which you are affiliated and that the buck stops there, but there are other aspects to most facilities that can impact the effectiveness of your work progression. These might include resources offered by the department (such as microscopes, animals, or cell culture facilities), established research stations, established networks with other institutions, or specific discounts at work-related suppliers and retailers.

Table 1.2: Institutional Resources

INSTITUTIONAL SERVICES AVAILABLE FOR: Xavier Marques Group					
Mouse Biology, Myopathy and Therapeutics					
TYPE OF SERVICE	DETAIL OF SERVICE	LOCATION	CONTACT	LINKED STAFF	COST
Computing	Provide IT support, programs, hardware/software, installations	Biology Department IT, Room 365	Service Desk, ext. 5468, IT@bio.uni.edu	Approval from X Marques for service over $100	Invoiced case-by-case
Learning Support	Provide assistance with design of courses and Moodle delivery	Uni Learning Center, Room 12	Service Desk, ext. 5490, LS@uni.edu	Postgraduates and staff may consult at any time	Free
Counseling	Personal counseling for all staff and students	Uni Welfare Center, ground floor	UWC, ext. 5423, UWC@uni.edu	Students and Staff receive 8 free consults per year.	Free
Grant Applications	Research grant application support, re: guidelines, structure, review	Research Office, building 5, room 341	RO, ext. 5434, RO@uni.edu	X Marques access	Free
Statistics	Statistical advice and analysis where required	Statistics Office, Math Department	Stats Office, ext. 5410, Stats@math.uni.edu	Postdoctoral and staff may set up consultancy	Consultation free, analysis invoiced case-by-case
Printing	Printing posters, theses, and large manuals, and photocopying service	University Printery, Central Services District	Uni Printery, ext. 5472, Uniprint@uni.edu	Accessable to everyone	Students = 10% discount, invoiced case-by-case
Equipment	Advice regarding locating and access to various equipment on campus	Faculty Manager, G48, Building 8	John Jones, ext. 5443 BioManage@uni.edu	Accessable to everyone	Free
Editing/Revision	Document language revision to perfect English standard	Language Services, Central Services District	Joanna Briggs, ext. 5436, languageservice@uni.edu	Accessable to everyone	Research students 15% discount, invoiced case-by-case

Business Strategies

Short-run business strategy

The predominant focus of this book is to promote effective research business management and development in the short-run scenario. In this case, the idea is to encourage the development of a functional research project and program that provides dividends for those participating, whilst encouraging a positive, inspiring, and productive workplace. This takes place day-to-day and project-to-project by **ensuring deliverables are met on the current project schedule** and meetings are kept, promoting overall group productivity in meeting set targets where possible, and troubleshooting issues as they arise.

Long-run business strategy

> **CASE STUDY: Long-run business strategy**
>
> Imagine you built your whole career on a single research goal, like developing a cure for a particular human disease, for which the genetic basis is not yet defined and the cure seems a long way off. Fast forward ten years and you have completed your PhD and postdoc, and you are 1–2 years into setting up your own research group, in which you are all focused on resolving a potential cure for this disease. The genetic basis is defined, and the overall mode of pathogenesis is described. There is a lot of funding for your research focus. Then, two years into your tenure as an independent group leader, the cure is discovered. Your whole research program is shot. You were partially involved in the cure that was developed, but the work is now rolled over into patentable technology under the control of a drug company that has sponsored the patent applications. Clinical trials are being planned. Suddenly, all your years of research focusing on this disease are redundant and you have to scramble to figure out what to do next. This is an example of overall bad strategy with regard to long-term career planning.

This case study emphasizes the need to always consider your long-run research plan. If you want to pursue a research career with any longevity to it, or at least nurture that as an option that you can choose, **you need to make strategic career choices** and think hard about what you want to do in the long run. You also need to think about where the next project is leading. This might mean that instead of just focusing on the one disease, you spin off other related projects depending on the technology that you develop. You might develop new technologies in your research pursuits that lead to fruitful collaborations, or you may be assigned to chair a key facility within your institution, which secures not only your job but also your research infrastructure. Being adaptable to change and embracing new directions is a key component of longevity in academic research careers.

BEING ADAPTABLE IS KEY TO CAREER LONGEVITY

<u>Defining your research goals</u>
Taken together, it is important to reflect on your short-term and long-term professional goals as a research academic. What do you hope to achieve? Make a list and keep track of how you are going to pursue these goals. How will you prioritize your time to maximize your advancement toward these goals? Table 1.3 outlines the type of spreadsheet that you might keep to track your current research goals, and to consider how they might grow with time. Exercise 1.3 at the end of this chapter encourages you to create your own table of research goals.

Table 1.3: Defining your research goals

Professional Goals: Xavier Marques			
Short-Term Goals	**Current State**	**Planned Approach**	**Deadline /ETA**
Develop therapy approach with B-Reg cells	underway	Postdoc/tech team	2 years
Establish smooth microinjection service	commencing	Technician/facility operator	1 year
Explore lead compounds for mimetic therapy	underway	Postdoc/tech team	2 years
Expand core concepts of intracellular signaling mechanisms	commencing	PhD student	3 years
Identify role of core compounds on dystrophies	underway	Postdoc	2 years
Long-Term Goals	**Current State**	**Planned Approach**	**Deadline /ETA**
Explore roles of potential therapeutics in different dystrophy models	anticipated	Expand from current postdoctoral project, linked with mimetic study	5-10 years
Expanded microinjection service offering external users, with regular collaborative publication output	commencing service this year	Offer collaboration and paid service	Ongoing
Develop patentable therapeutic and expand to industry-funded spin-off	starting	Roll over from current lead compound research	5-10 years
Apply advanced knowledge of intracellular pathways to develop more specific therapeutic approaches	anticipated	Expansion beyond current PhD project, core research	5-15 years

Marketing

You may not consider yourself an expert marketer as a scientist, but marketing yourself and your work is a key part of being a professional academic researcher. These days there are a multitude of ways to market yourself, and some of the most talented and popular researchers are strong science communicators.

Finding ways to **publicize your research** can significantly impact your successes in attaining research funding, attracting talented staff and students, and aiding collaboration development. Developing your own group website is a sure-fire way to improve your marketing, by developing your own personal "brand" so to speak. Developing educational videos related to your technology, or running courses to train other researchers — through your institution or through larger organizations — can equally expand your network. Presenting regularly at conferences is the more traditional way to promote yourself, but internet-based promotion through personal and institutional webpages, Twitter, Facebook, blogs, and other social media will also only enhance your research profile. These approaches are discussed further in chapter 5 and appendix 1.

Key Players in the Research Business
The Group Leader
If you are running an academic research group, the chances are that you've gotten there by being an amazing researcher in your own right. You have likely published significantly in your field of interest and now you are in the position that you can expand your project operations and increase your productivity with a more senior, and likely more secure, position.

It is likely that your official leadership experience is minimal when you begin this role and that you are probably overwhelmed with the amount of work you need to get done: teaching, institutional bureaucracy, project development and funding acquisition, as well as managing your group and projects, and any collaborations that you may have established. You do not need any drama related to your staff and students, and you want a smooth operation that involves minimum effort and maximum productivity.

The Postdoctoral Fellow (postdoc)

Postdoctoral fellows come with a broad range of attitudes, from those who are motivated and keen to get started building their own research empire immediately to those who simply cannot think of anything better to do professionally and are there for the pay. No matter what category, postdocs need to be able to navigate the research environment to deliver the appropriate project outcomes. Postdocs often need to help in the supervision of junior group members, and may or may not contribute to teaching within the institution. Postdocs are often required to write funding and grant applications as par for the course.

The PhD student/doctoral student

The predominant focus of all PhD students should be on the project and degree program that is to be completed in order to achieve their award. This should go hand in hand with a solid plan for publication during the candidature. Doctoral students typically operate on scholarship funds, and they usually prove to be dedicated researchers that most typically see projects through to completion. Thus, they represent valuable members of the group who may stay on and continue to maximize output, if given the chance after their degree program is over. Solid PhD experiences can promote lifelong professional collaborations that can span entire careers.

PhD students require various levels of guidance, depending on the individual. Extra peripheral training courses and professional development should be encouraged, including training to operate resources in the group and to present their work well, both orally and in writing. Recruiting a doctoral student is a savvy way to secure someone to run a project to completion — that is, someone who has a vested interest and dedication to achieve a successful outcome. However, PhD students do require supervision and guidance, and thus good management.

The Masters or Honors research student

These junior graduate students typically perform a short (usually up to one year) research project that should represent a tidy investigation, which is designed by senior researchers and focused on generating clear datasets that the student can assess and present in a thesis. This may or may not generate a full publication, but it is of great benefit if the student can contribute to a larger study and get their name on a publication. These students can often slip directly into the PhD program after graduating, so training them well and supporting them through their studies can represent a savvy investment in staff development.

The Technician

Technicians can form the backbone of a research group, operating as group coordinators that are focused predominantly on the smooth running of your group and effective organization on a practical level. Their level of responsibility will vary from person to person, but some technicians can know the work just as intimately as the group leader. Whilst students and postdoctoral fellows tend to move around every few years following projects, fellowships, and grants, the technician can be retained for multiple projects, and thus get to know your operation to make it run smoothly. Technicians are well worth the investment, particularly in a larger group with several funding streams.

The Intern/Undergraduate/Summer student

Undergraduates typically enter the research group to complete short summer placements (usually of 5-10 weeks) to gain some practical research experience. They can represent a useful resource for the group, but they often require a lot of training during their placement. The upshot is that, oftentimes, undergraduate trainees are looking for somewhere to continue on for their postgraduate research training, and a long-term relationship can evolve between the group and the undergraduate trainee. Similarly, training undergraduates and directly mentoring junior researchers are essential skills for postdoctoral fellows and PhD students; indeed, these are skills which they will require when moving forward in their career.

Clearly staff represent a key valuable component of your research group, and they require a solid leader to ensure that the group operates effectively. Different strategies for managing the staff in your group are discussed next, in chapter 2.

END OF CHAPTER 1 SUMMARY
Running a Research Business

In this chapter we have introduced the research business model, with a focus on establishing your business structure and plan. Some of the issues discussed have included:

- You must wear many hats as a research group leader.

- Proactively developing a successful and funded research group is key to success.

- Consider the value that you are adding to your institution as a researcher, and capitalize on that.

- Establishing an outline of your current research business can help you to consider your overall "business structure."

- Institutional services and facilities can enhance your business and should be recorded to promote accessibility and awareness for you and your group.

- It is important to consider how you see your group progressing in the short-run and long-run, so that you can plan your research accordingly.

- Marketing your research is an important way to expand your profile. Innovation in communication is encouraged.

- There are several key players in the research business, including staff and students. Refer back to the overview in the section above.

REFLECT

1: What special value are you adding to your department/institution? What makes you worth employment?

2: Is your research group running effectively and delivering "value-added" to your institution?

3: Do you have defined goals and deliverables that you are aiming to achieve in your research career? Is your group research focus clear?

4: Do you market or publicize your research in the public digital or media space? Have you developed a research "brand"?

EXERCISES

EXERCISE 1.1: MAP OUT YOUR OWN RESEARCH BUSINESS

Create a spreadsheet or document with the following titles and complete this exercise to establish how your research business is structured. See the example from table 1.1. You can download this example and a blank spreadsheet from the Practical Academic website.

RESEARCH GROUP OF: Group leader name					
Group focus					
DATES	INPUT	GOAL	PROCESS	YIELD	STAFF

EXERCISE 1.2: OUTLINE YOUR INSTITUTIONAL SERVICES

Create a spreadsheet or document with a list outlining details of your institutional services. This exercise will help you to create a source reference for your group, so that they can be aware of the support or services that are available to them. The example shown in this chapter (table 1.2) and this blank spreadsheet can be downloaded from the Practical Academic website.

EXERCISE 1.2: OUTLINE YOUR INSTITUTIONAL SERVICES

INSTITUTIONAL SERVICES AVAILABLE FOR: Group Name

Group Focus

TYPE OF SERVICE	DETAIL OF SERVICE	LOCATION	CONTACT	LINKED STAFF	COST

EXERCISE 1.3: DEFINING YOUR RESEARCH GOALS

Reflect on your short-term and long-term professional goals as a research academic. What do you hope to achieve? Make a list and keep track of how you are going to pursue these goals. How will you prioritize your time to maximize your advancement toward these goals? You can download an example and this blank spreadsheet from the Practical Academic website.

Professional Goals: YOUR NAME			
Short-Term Goals	**Current State**	**Planned Approach**	**Deadline /ETA**
Long-Term Goals	**Current State**	**Planned Approach**	**Deadline /ETA**

EXERCISE 1.4: GOING FURTHER — DEPARTMENTAL STRUCTURE

Make a mind map of your departmental/institutional structure and consider where you fit in that map. Structure your map from the top down, which is the standard structure of academic research institutions, and consider the point of view of the people filling each individual role. Identify how you would like to progress your career through the institutional structure over the next ten years. Would you like to change institutions?

Chapter 1 — Downloadable Materials
Download from www.practicalacademic.com

- Table and exercise 1.1: "Map out your own research business" — Excel file containing a blank template and the example provided in chapter 1.
- Table and exercise 1.2: "Outline your institutional services" — Excel file containing a blank template and the example provided in chapter 1.
- Table and exercise 1.3: "Defining your research goals" — Excel file containing a blank template and the example provided in chapter 1.

REFERENCES – Chapter 1

Betz, F. (2007) Strategic business models. *Engineering Management Journal*, 14(1) 21-27.

CHAPTER 2

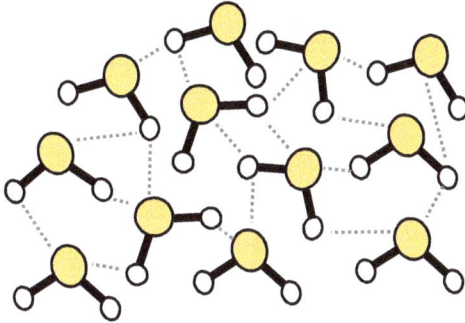

People and Management Structure

Projects in academia tend to vary according to discipline, topic, experimental approach, and duration. However, all academic institutions generally employ group leaders based on their academic credibility as an expert in their field to pursue their own line of research. Most academic leaders lack formal training in management, and those that excel often tend to do so due to natural instinct and luck. The way that these leaders structure their groups tends to vary dramatically, but there are several common models, which are shown here for your reflection.

Group Structure Models

In this chapter, several models are put forward relating to different approaches that group leaders might take to manage their research groups. These models are purely based on observations gained from experience working in several research institutions. Reflecting on how effectively these models worked for different research leaders, staff, and students makes it obvious that different styles suit different people. Styles can be dictated by schedules, projects already underway, institutional frameworks, funding, and — most importantly — different personalities and personal preferences.

It is clear is that groups are well served by taking time out to consider whether the approaches that are currently in place are working. In doing so, alternative styles of group organization should be considered. In reading this chapter, you should be thinking about what aspects of these models appeal to you, and which approaches are you currently taking in managing your people and projects. Being proactive in considering the style that suits you and your group will ultimately pay dividends by raising your awareness of your own approach to management. Once you can identify how you are managing, you will have a foundation from which to improve your practice.

Remember: these are models, and thus generalizations, of approaches that group leaders may take in their group management. A great number of variations on these models are very likely, and they are worth discussing further.

Model One: THE INDIVIDUAL PATH

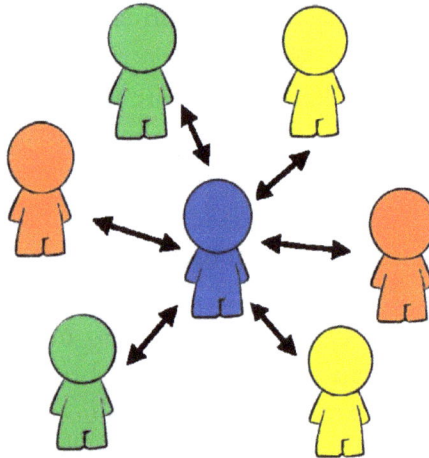

The INDIVIDUAL PATH is a typical approach taken by many supervisors. It aims to promote **individual focus** on an **independent project** that may or may not relate to other projects currently underway in the group. In this model the group leader assigns each individual in the group their own project focus. This does not negate the idea of collaboration within the group, but nor does it promote it.

This approach can include collaboration with others, but typically the person assigned to the project is the manager, lab monkey, data cruncher, writer, *and* presenter of that project.

ADVANTAGES: This approach can be empowering if you manage to get your project off the ground, if the stars align, and if all of your experiments work. Successful participants in this style of group organization tend to be the kind of people that work themselves to the bone until the project succeeds. Individual researchers can develop a strong capacity to work independently, which is likely to be useful in future projects if they end up in another group that manages projects this way. Furthermore, the group leader can maintain a close watch on all of the projects in this setting.

DISADVANTAGES: This model relies heavily on chance. People taking the individual path in their academic work tend to be isolated, and they often get depressed when they hit stumbling blocks and lack people with whom to discuss ideas. Individual researchers do not tend to collaborate very much on their work, and thus fail to gain such skills. This deprives junior researchers, in particular, of the opportunity to develop collaborative experience. This model can result in significant group conflicts due to competing interests and resource sharing in a survival-of-the-fittest environment. Individual researchers can form alliances in the group, and the more manipulative group members can thrive in this environment. The group leader needs to put more time and effort into this approach, if it is to be done well, given the individual attention required for each individual project/group member. Nonetheless, some projects do tend to garner more attention than others, and those working on the less attractive/productive project foci tend to receive less support.

Model Two: THE TEAM APPROACH

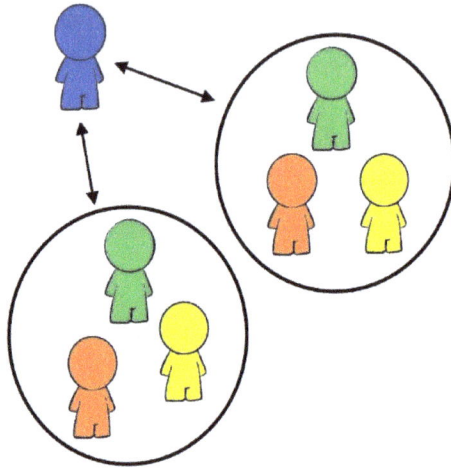

The TEAM APPROACH is typically demonstrated by the participation of one Professor, one Postdoc, one PhD student, and one technician. Each participant has a particular goal in mind. The professor wants project completion and publication, the postdoc and PhD students want publications to promote their careers, and the technician assists wherever required.

ADVANTAGES: Nobody is competing directly against each other in this model. This requires the respect of each participant, but ultimately it can be a winning combination for all involved. The PhD student benefits from the direct input of two more senior researchers, one of which is directly present for the entire project. Thus they have direct and detailed guidance regarding their work and career. They also learn how to work better in a team, and their work is propelled forward by the team approach. The student can still be the one to make the decisions on the main focus of their thesis work, but they can also contribute to other projects that can be considered "side" or "secondary" projects, thus expanding their publication output significantly.

All parties benefit from the collaborative approach to planning their work and completing it as a team. This approach is considered by many to be the most effective way to run an academic research laboratory, and it minimizes the time that the group leader is required to "manage" week to week, given the level of delegation that can be given to the team and the reduced requirement for individual meetings. See the case study in chapter 6 — "the team approach."

DISADVANTAGES: Friction may occur where the postdoc and PhD student clash when it comes time to publish. However, this can be addressed if they can discuss this issue and come to an agreement from the outset, and then continue an open dialogue regarding publication rights as the project evolves. Furthermore, team members may leave or join in the middle of a project, and this may lead to clashes — particularly when a new postdoc comes in mid-project. Such clashes can be about who takes the lead, especially when a PhD student might regard themselves as lead on a particular project. More communication is required to ensure that all participants are on board, and sometimes friction can arise when those team members who preferred individual projects are forced to work in a team, or when one team member fails to contribute equally.

Model Three: THE 2IC APPROACH

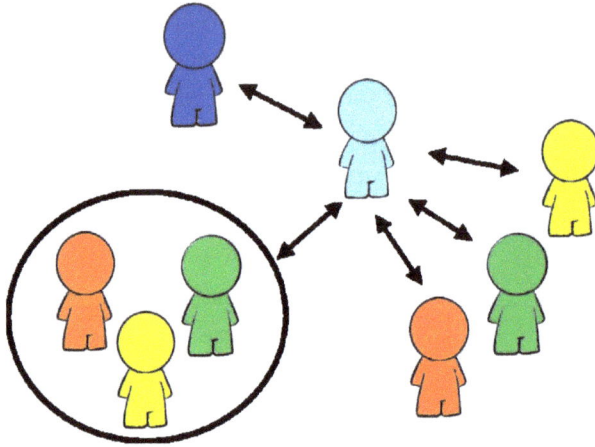

The Second-In-Charge/Second-In-Command (2IC) approach looks a lot like the team model, but ultimately the work is driven by the person who takes the role of "second in command." This is typically someone who has been working in the group since grad school, and who — for all intents and purposes — is in charge of the scientific direction of the group.

ADVANTAGES: This structure is often seen in big groups with a figurehead lab leader who might be so high up that they focus most of their time on other duties, like heading an institute, contributing to government activities relating to science, or holding several groups between different institutes. The 2IC can effectively run research programs from this position, modeling any of the other structures as if they were the head, and this approach is proven to be effective. This model can work well if both the big boss and the 2IC perform their roles well.

DISADVANTAGES: There can be issues with postdocs, in particular those who may have gone to work with the "big name" science leader, who resent the fact that they may be forced to work under someone who is not necessarily that much more experienced than they are. The 2IC may also be trusted due to their many years of service, but may lack people management skills.

CASE STUDY: 2IC approach

A postdoctoral fellow is recruited to a highly-regarded laboratory from abroad, having secured a competitive fellowship. The fellowship brings substantial research funding with it. After arriving, the postdoc realizes that they will have very little to do with the lab leader and is expected to run their own program with minimal input. In fact, this postdoc is required to work directly with the 2IC senior fellow who has been in the group for many years, but who does not necessarily have a very good professional ethic. The 2IC does not make much effort to direct the research work and treats the new postdoc rather flippantly in group meetings and over lunch. The postdoc works very hard, but is required to list the name of the 2IC on any work that they present — at conferences, in seminars, and for any papers that come out. After two years, the postdoc feels demoralized and disillusioned with research, and decides to leave the bench, instead of pursuing research work further.

This is an example of what can happen when the 2IC model goes wrong. Can you identify the source of the problem?

Model Four: THE COMPETITION APPROACH

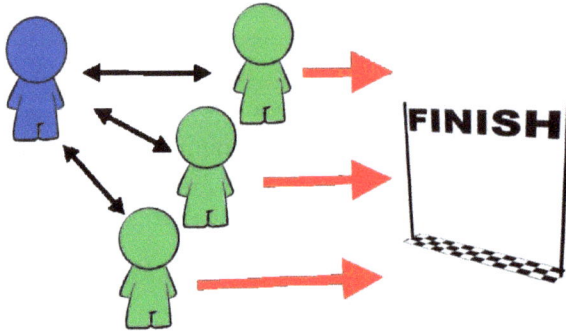

The COMPETITION APPROACH is typically found in very well-funded groups that are under intense time pressure to publish or perish. Group leaders have been known to employ many postdocs, often placing multiple independent postdocs on one project focus and encouraging them each to work alone on the project, thus pitting them against each other in a race to complete first. A variation of this may be where peers are working on different projects but competing for the same accolades.

ADVANTAGES: This structure is sometimes seen in big groups that have a lot of money and which postdocs seek out to experience research at the cutting edge. Often the group does demonstrate good progress in their field, but typically at the expense of staff wellbeing.

DISADVANTAGES: The competition culture in research can bring out the worst in people, resulting in disheartened or disillusioned researchers who could have an incredible and positive impact on scientific advancement, but who sadly often leave the profession. This model demonstrates a cut-throat approach to group management, which promotes all sorts of misconduct in researchers participating in this setting. Researchers can often waste years of their life pursuing outcomes that either do not exist or which they are out-published on.

The few that make it in this setting tend to perpetuate the model. There is a lot of misery and bad feeling associated with this model, and the workplace tends to be defined by arguments and disrespect. There is little evidence of mentorship from more experienced researchers in groups that follow this model. Furthermore, groups that operate this way tend to demonstrate more plagiarism and dodging of data than other models. This approach holds more chance of researchers being blamed for unreliable data generation, given more pressure to produce.

CASE STUDY: The competition approach

In a competitive laboratory, where one postdoc had bowed out after six months of chasing rainbows on a particular project focus, a second postdoc picked it up. After a short time, they provided beautiful results clearly demonstrating the physiological mechanism that was the focus of the investigation. Only thanks to one savvy group leader, who asked to investigate the raw data, did they discover that this shameless postdoc was falsifying data and manipulating data points to fit their hypothesis.

<u>Model Five: THE LONE WOLF</u>

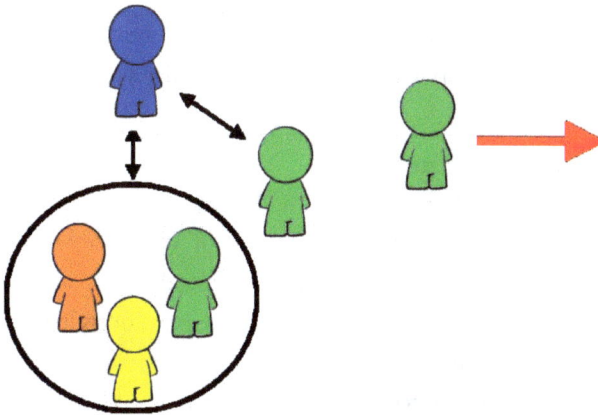

The LONE WOLF researcher model evolves when a researcher ends up in a group where either the group leader has very little involvement in the work being done, or an independent researcher comes to utilize the group's resources for a defined project. This model can also evolve due to self-exclusion or social exclusion in a particular group. Lone wolves sometimes arise due to their natural personality, whereby the group leader accepts that the researcher works independently in their group but is not required to liaise regularly with the group leader regarding scientific direction; typically, this happens when the researcher has demonstrated excellence in research productivity, and when they have their own research fellowship and funding. People have been known to complete years in this environment, creating, directing, completing, and publishing work with which their group leaders have minimal involvement — until publication, of course.

ADVANTAGES: This approach can work well for specific personality types, or for researchers at particularly self-reliant stages of their work. This might operate in tandem with other approaches to group management, or it might not.

DISADVANTAGES: Lone wolves can end up spinning around in circles and getting little done if they fail to establish the right approaches to promote productivity. They can benefit from regular check-ins with their group leader, to create accountability in their work.

CASE STUDY: The lone wolf

A quiet and hardworking research fellow from a non-English speaking background pursued independent work separate to the rest of a cohesive group. He had little interaction with his supervisor and, when he hit stumbling blocks in his productivity, he never chose to ask for assistance. The group leader struggled to engage and support him. After two years, he had little to show for his time, despite working long hours in the lab.

In this case study, the fellow could have benefited from peripheral support from the host institution in the form of English-language groups. The group leader could have established regular 'check-in' meetings to keep track of their work. Similarly, the research group could have made more effort to include him, both professionally and socially.

<u>Model Six: THE SMALL GROUP</u>

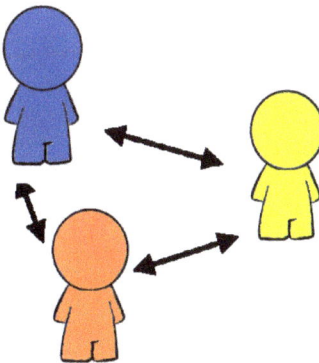

The SMALL GROUP is most common when you start out as a group leader in a new academic position, especially when you are transitioning from being a Research Fellow. In such cases, you are likely to be running a small lab. This might be operated on a tiny budget with only junior postgraduate students or one technician working for you, and you might still be spending all of your spare waking hours continuing to do the research yourself in order to set up your new independent research investigation focus. This model is where most academic researchers start out. If you are good at securing grant money or your department provides outstanding start-up expenses, you may even be able to get postdoctoral fellows or PhD students in your service early on.

ADVANTAGES: When small labs are new, they tend to be fresh and exciting places to be, with great new research ideas and a very motivated group leader. This works particularly well when the small group is adjunct to a larger group where the young group leader was originally working as a research fellow, so that the new group can develop its own independent focus with the support and facilities of the adjunct group. This may also take the guise of collaborative support. The small group can involve a very tight-knit team who are working together to achieve common goals. This type of setting can set up a research student to become a 2IC as they pass through their research training.

DISADVANTAGES: Much pressure is on all members to produce in the small group, and oftentimes the young group leader takes a while to come to terms with the extra responsibilities and expectations that come with transitioning from a research-focused role to a tenured position. Distractions from developing teaching resources, managing departmental responsibilities, and developing research — all whilst establishing new facilities and writing grants — make the role overwhelming for many new group leaders.

CASE STUDY: The small group

A clinician was appointed an adjunct research position at a highly regarded tertiary institution. She was seeking an opportunity to extend her clinical knowledge by developing some core research focused on a disease she often treated in the clinic. She associated herself closely with a larger established research group in the field, and she recruited a technician and a PhD student to work with her.

The clinician gave direct and individual attention to her researchers when she met with them each week, and these researchers also benefited from being associated with the team in the affiliated larger research group: they benefited professionally and socially, and through access to more established resources. Although the group remained small, it benefited from the combined clinical and research access, and it promoted innovative publications in their field.

Model Seven: STUDENTS ONLY

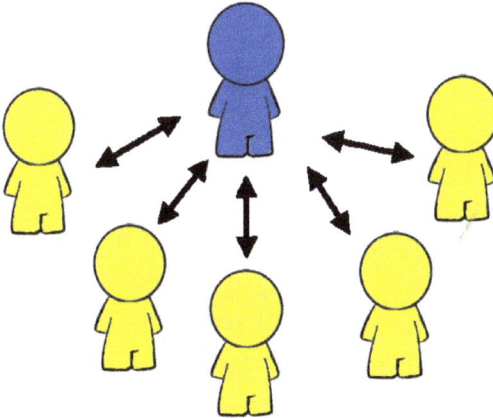

The STUDENTS ONLY model often comes about from the small lab model, or in periods of research during which funding is low. In this scenario the research will be run on a budget, and little will remain for research-assistant support and postdoctoral scientists. A large number of Masters and PhD students may be accepted to one group, in which the only staff member is the group leader.

ADVANTAGES: This model represents one way to maintain research activity whilst going through a transient funding slump, whether mid career or just starting a new group. It can work well if the students are managed effectively, and if they gel socially within the research space. Continuing the work this way can pave the way for innovation into many research streams and funding opportunities, particularly when co-supervising students with other group leaders in a collaborative endeavor.

DISADVANTAGES: Students can become quite competitive in this environment when they are not managed well. Group leaders have to put in a lot of work to manage their individual students/projects. Students can reach stopping points in their research if the lack of funding in the group prohibits access to resources.

CASE STUDY: Students only

A young group leader accompanied his old friend and colleague in establishing a new research focus in a major university department. He had start-up funding from the department upon arrival, but he missed out on the competitive government research grant that year. He gradually recruited Honors students to join his group and had them mostly performing computer modeling and theoretical analyses in his field, so he could achieve research outcomes on a budget. As the students graduated and commenced publishing their work, more funding support came and many of them stayed on to complete PhDs and more. It represented an effective way for the group leader to get his research group off the ground, given the circumstances. Most importantly, they all loved working together, becoming fast friends and future professional collaborators.

Changes in Group Structure

Changes in group structure obviously take place as your career evolves. Most new group leaders start out with a simple "small lab" structure, or as an adjunct leader to a section head. Far and away the easiest way to get established is to develop your own independent research focus under the umbrella of a larger group, and then step sideways within the institution into your own tenure-track position. Nonetheless, sometimes opportunity knocks from new institutions and perhaps you will have to take the opportunity for whatever reason: personal reasons, location, an excellent package offer, outstanding facilities to accommodate your research, amazing funding, or job security.

Joining a New Lab

For junior researchers, once they enter your group it is worth clarifying at the outset who they will be working with. Similarly, how the lab is managed should be clearly explained to students and postdocs at the outset. It is important to clarify and promote awareness of all of these styles of lab management, and how they can impact people in any role in the hierarchy. The approach you choose can have a significant impact on the group's individual productivity, social life, and general happiness in the workplace. It is good to reflect on these models and gauge which approach fits best for your group and the people that are a part of it. Which style seems to best fit your approach to research?

What Impacts the Style of Management?

Personality and experience play a big role in defining the type of research group that you establish in the academic sector, whereas productivity and/or the topic of research focus tend to play key roles in the process of being recruited as an academic group leader. Your political persuasion, dietary choices, wardrobe choices, and hair cut rarely play a role in securing a research leader position; similarly, you do not necessarily have to have any significant training in project management or people management. That said, consider pursuing whatever courses are available to you at your institution to **acquire extra management skills** in these areas. They will strengthen you as an academic leader. It is important to remember that you never stop learning, and taking on a management role in this setting is only the first step in this stage of your career progression. You have to think smartly about how you are going to create a productive research group over the next ten years to consolidate yourself as an independent and quality research leader. Next we will go on to discuss human capital.

END OF CHAPTER 2 SUMMARY
People and Management Structure

In this chapter we have discussed the pros and cons of different group structure models that are commonly seen in academic experimental research groups. These structures are:

- The individual path — separately managing each group member under your supervision.
- The team approach — establishing collaborative project teams within your group.
- The 2IC approach — designating a second in command to lead in your absence.
- The competition approach — offering the same project to multiple people in a race to complete.
- The lone wolf — people in your group working independently with little oversight.
- The small group — one to two students or staff working to establish a central research focus.
- Students only — supervision of multiple individual student projects in tandem.

REFLECT

1: What approach are you taking to managing your research group? Does it fit into the models described? How is it different? What are the pros and cons to your current group management structure?

2: How has the structure of your research group impacted your time and productivity? How do you feel you could improve on this?

EXERCISE 2.1: YOUR RESEARCH TEAM
Make a summary of each person in your research team and outline how you have established their management relative to the different

approaches outlined in this chapter. Consider how this is working out for each of your group members in your current research set-up.

Group member	Position/contract duration	Type of management structure

EXERCISE 2.2: SUMMARY TABLE — Management Approaches
How would you rate the different models of group leadership
regarding the following characteristics? Rate from 1 to 10.

Management style	Support	Productivity	Happiness	Cohesion	Efficiency	Sociability	Cross-disciplinary	Time commitment
Team model								
Individual path								
2IC								
Competition model								
Lone wolf								
Small group								
Students only								

This table outlines the models described in this chapter with reference
to various characteristics of the group dynamic. The style of group
management can have tremendous impacts on overall group
dynamics and the experiences each member has. Many characteristics
depend heavily on how the group is managed by the group leader.

Chapter 2 — Downloadable Materials
Download from www.practicalacademic.com

- Exercise 2.1: "Your research team" — Excel file containing a blank template and an example answer.
- Exercise 2.2: "Management approaches" summary table — Excel file outlining the characteristics of the seven group management models put forward in chapter 2.

CHAPTER 3

Human Capital

Institutional academics spend a lot of time critiquing each other. This is a function of the academic sector, defined by "peer review" and the pursuit of academic quality. Mentors and leaders can be harsh and soul-crushing task masters that demand excellence. It is a jungle out there. But ... your group does not have to operate this way to function well, and if you manage your group well, you will retain outstanding staff that see your group as a haven of tranquility where they can pursue academic excellence between bouts of "peer review."

If you consider your group to be your business in a bustling marketplace, you would be foolish to operate in a way that results in unhappy staff. When your workers are discontented, you often see poor performance, reduced staff retention, and high turnover, and thus higher costs for the business. In the academic research business, a group member's knowledge-base and skillset are critical in the effective pursuit of innovation.

Your postdocs and postgraduate students are potentially your closest allies and best future collaborators in the research world. Similarly your technician, whom you have trained to operate the laboratory and its requirements over (usually) years of their lives, is going to perform any task related to your work more effectively than any newcomer, whether they arrive with a shining CV or not.

Retention of skilled staff makes your life as a group leader and/or project manager (see section 2 for the differentiation of these roles) easier, particularly if you have bothered to take the time to train them well and to promote a positive working environment within your group.

Recruiting

Getting the right person for the job is a critical skill in the research sector. Typically you will have a window within which to recruit in order to meet the expectations of the funding body and, unless you've already got someone in the wings, finding the right candidate can be a nightmare. Various sources of skilled staff are noted here.

Sources of New Staff

Someone you know

One of the easiest ways to fill a funded position in your research group, with someone that knows what they are doing, is to employ someone you know. In particular, someone who already has experience in your group or department, or who has participated in collaborative research with you, will be able to pick up a project quickly and hit the ground running. They will not necessarily require induction or training, and they likely already know your team.

This type of recruitment often involves taking on board a summer student, a recently-graduated research student, or someone whose contract is up in an associated group in the department.

Word of mouth

Good researchers tend to maintain positive professional networks. So, when they have funding to employ someone, they can spread the word that they are looking and often a peer will recommend someone they know. This is a very popular approach to recruiting postdoctoral fellows or technicians for new projects. Similarly, if one of your colleagues is teaching a particularly bright student, they may come recommended to you for a Masters project.

<u>Direct recommendation</u>

If someone in your group is looking to expand their professional experience, it is entirely acceptable to help them get their foot in the door of a new group by providing a professional recommendation to another group's group leader. This is a powerful way for junior researchers to pursue their next research position, because a positive reference is indicative of having performed well and successfully in their most recent position.

<u>Direct inquiry</u>

Potential new researchers will often approach group leaders directly to inquire about the possibility of working in their team. People that make the effort to inquire should not be overlooked, because they usually bring with them an intrinsic motivation to succeed in the role. Direct inquiry is stronger when backed up by good references, but it is best when the candidate brings some interest in the work and knowledge of the group's recent work.

<u>Job advertisement</u>

Although most institutions are required by law to advertise professional positions prior to filling a role, many advertised positions are already coveted by someone who has a prior knowledge of the group, as outlined above. Nonetheless, competitive calls can be made through advertising a position that is available.

Advertising a role can attract a strong field, but that means you can establish your own database of potential employees, should a position become available in your group in the future. It is certainly worth interviewing the most professionally attractive candidates, even if you already have someone in mind.

Sources of Salary for New Staff

Staff will be employed from a range of financial sources, including the following ones.

Employing from a grant

Research grants can provide for research staff to facilitate the project outcomes. Typically employed on 1–5 year contracts, recipients of such grants can include research assistants, postdocs, graduate students, and consultants.

Employing from a fellowship/scholarship

Postgraduates and postdocs can often bring their own funding to a project. This funding covers not only their salary, but can carry peripheral support for the fellow, including research expenses, travel expenses, equipment, and relocation costs. Acquiring fellows that bring their own funding is an asset to the group, relieving the pressure on the research grants to be the main source of funds. Similarly, a postgraduate or postdoctoral fellow who sources their own funding demonstrates their independence and capacity to drive their own research, and to acquire their own funds.

Institutionally-funded technician

If you are running a core facility for your department as part of your tenure, there is a good chance that the department will foot the costs of the technical staff to operate that facility day-to-day. This can represent a long-term prospect when it comes to project development and completion, providing a reliable participant in your research group. It is very likely that the technical staff under your guidance would welcome the opportunity to contribute to your research in order to get their names on published papers to promote their career productivity, whilst also enjoying the stability that is brought by working as a permanent technician. This is discussed further in chapter 8.

Employing from soft money

Most groups have the capacity to build up soft money as a slush fund. Sometimes this can be used to hire staff. This can often take the guise of short-term appointments, top-up scholarships, a support position that is only for a few hours per week, or consultants.

Collaboration employees

Collaborative employees might fall into the category of visiting scholars, joint supervision staff, or students. Collaborative employees might be independently funded or co-funded, and they typically create or continue collaborative links between two individual research groups.

Volunteers and unpaid students

Short-term volunteers and unpaid interns typically join research groups during professional experience training in the later stages of their undergraduate training. They may or may not be able to bring a short-term stipend or research expenses.

Recruitment Interviews

At the commencement of assigning anyone to a contract in your group, an interview is necessary to determine their fit with the role.

Things to look out for in recruiting:

- self initiative,
- alignment of background with expectations of role,
- ability to deliver (e.g. availability, experience in sector),
- potential cohesiveness with established group,
- capacity to complete the contract in the assigned period, and
- agreement on the work approaches taken in the group.

Interview Types

Interviewing new candidates can be done in a range of styles that may depend on your institution. Some examples of interview approaches are shown here.

Small response, low level position, informal interview

When only a few candidates (e.g. 5–15) apply for a junior research position, usually an informal interview is sufficient. This occurs between the group leader, the candidate, and sometimes a postdoctoral fellow, if the latter is to be involved in co-supervising the new recruit in question.

Medium-to-high response, panel interview

Panel interviews are typically used for calls that result in larger responses, and which are made to recruit a research fellow or senior technician on a grant or in an institutional support position.

Official departmental interviews

Official institutional recruitments may involve several rounds of interviewing, often involving key candidates making presentations on their work to the department or to a selection board. Each institution has its own approach to official appointments, but they typically happen in response to open calls, and in a completely unbiased process, without favoring any individual.

Video interviews (e.g. Skype)

Skype interviews are becoming more standard in the recruitment process, particularly for institutions that are recruiting staff from an international talent pool. This opens up the field to secure the most qualified candidate for the position, and it typically includes a panel of interviewers.

Telephone interviews

Telephone interviews represent a good way to have a one-on-one interview to hire junior staff, or to have a pre-formal interview discussion about the role. They tend to last no longer than half an hour, and they give the employer a chance to gauge whether a candidate would be a good fit for the role.

The Role and Value of Each Group Member

Each group member can offer "value added," and it is important that all members appreciate the value each person brings to the work.

Group leader

The group leader's job is to set the focus and direction for the whole research group, providing a facility or resource set-up in which the group can pursue their research work. This typically includes access to an office desk and computer, where group members can do the written part of their research, and access to a resource area, be it a lab bench, facility, research station, field station, hospital resources, or other types of resources appropriate to your work.

The group leader is responsible for ensuring that the whole group operates effectively to achieve their research goals — typically publication in peer-reviewed journals, or the development of intellectual-property or of resources, the latter of which enables the group to innovate further.

A central role of the group leader is to establish the group structure and general culture that it incorporates. Similarly, the group leader's encouragement and promotion of career development for group members, and their provision of mentoring, is critical.

Postdoctoral fellows (postdocs)

Postdoctoral fellows carry out research investigations, typically in an independent fashion with the mentorship or guidance of the group leader. Usually this professional training period represents preparation for faculty positions through the acquisition of more research skills and practice in running research projects.

Postdocs are expected to deliver publications in peer-reviewed journals and to present their work in collegiate forums, like workshops, whilst also potentially supervising or mentoring junior group members. Postdocs are typically funded through their own fellowships, or employed on research grants.

Doctoral (PhD) students

Doctoral students join research groups to develop skills in being able to run an extensive and detailed research project, to deliver peer-reviewed journal articles on the topic, and to write and submit a thesis outlining the entire research study, which may be a compilation of individual studies on one particular overall focus.

Doctoral students are key components of the engine that keeps research groups going. They are typically funded through scholarship schemes separate to research grants, and they sometimes come with research funding to complement their stipend. Doctoral students tend to be very driven to succeed and mostly guaranteed to remain committed to the research until it is done.

Whilst doctoral studies usually take a minimum of three years, these students do often remain committed to a research project or group for many years to come, so as to maximize their output. Often times, doctoral students can come from previous work in the same group as a Master's student, technician, or undergraduate trainee.

Master's / graduate research students

The pre-PhD research training varies between institutions and countries, but usually involves 1-2 years of research, culminating in a short thesis. Master's/graduate research students can add significant value to your group's productivity. Master's students usually come directly from undergraduate courses, and thus have minimal research experience in the field. It is recommended to place a Master's student under the direct mentorship of a more senior scientist, and to encourage clear outlines of their work prior to commencement.

Master's projects provide a good opportunity for the development of simple research projects that might be awaiting a set of hands to complete, and which perhaps have no set deadline for completion. If the work goes well, Master's projects can often roll over into grant and fellowship applications. This facilitates the student staying on in the group to carry on the work, and to progress the investigation further. These students may opt to follow on into PhD training, particularly if their work goes well.

Undergraduate trainees

Undergraduate trainees often come to research groups in their final summer break before graduating from an undergraduate course. They often do this to try out practical research as a professional experience, to gauge their interest in the work. Although they usually conduct very simple investigations, usually directly assisting a more senior researcher (a postdoctoral fellow or PhD student), undergraduate trainees can move directly into postgraduate studies with the same group if the training period goes well and they are able to secure funding to proceed in their practical research education.

Technicians / research assistants

Technicians and research assistants may come from a range of professional backgrounds depending on the tasks that they are completing in their role in the research setting. Technicians are usually employed directly from research grants, or in some cases as departmental technicians, assigned to work under specific group leaders to provide a service to the department. Examples of such services are: tissue culture facility, animal production, microscopy, FACS analysis, or managing other specialized equipment.

Induction of New Researchers — Students and Staff

When welcoming new staff or students that are joining your group and department, you should cover several things. These things can be often overlooked in the informal research environment.

1. Safety.
2. Ordering, budgeting, and data management.
3. Access to equipment, reagents, and consumables.
4. The layout and use of specific work spaces and operating areas.
5. Group organization, schedules, and meetings.

6. General expectations of the person's role within the group.
7. Specifics of any mentoring agreement.
8. Expectations of new staff member.
9. Day-to-day expectations, schedule, and plan.
10. Project background information and the logistics for their role in the work.

Supervision / Leadership of Staff

Mentorship

Good institutions should promote mentorship training, but sadly many do not. Similarly, group leaders often either scoff at the potential "value-added" or simply lack the time or mental space to accrue any more training, given their requirements to teach, manage, mentor, write grants, and so on, as is the case in modern competitive academia. Thus, the amount of mentorship offered by any given project or group leader varies widely, as do styles and the consideration given to the professional relationship.

It is key to remember the responsibility that you take on when agreeing to be the mentor of a research student or postdoctoral fellow. Mentorship, at its core, is when you agree to **guide a less experienced or less knowledgeable person, and to do so in a process involving extensive communication.** Mentorship involves the transmission of knowledge, in addition to professional social capital and psychosocial support, relevant to work, career or professional development (Bozeman and Feeney, 2015). Many group leaders mentor by the "school of hard knocks." This means they expect their graduate student to "rise up" and stand to the challenge of the project put in front of them. Other group leaders micromanage and provide no freedom to explore the mentee's professional pursuit.

You should meet to discuss the expectations from both parties prior to commencing any new professional relationship or research project. Doing this before the new appointment begins, facilitates definition of expectations at the outset. Both parties should read any institutional guidelines on expectations from this working relationship.

A mentoring checklist is shown in table 3.1 to provide a guideline of what to discuss in a mentoring meeting. You can choose to create a formal or informal agreement, but you should write down and each keep a copy of what you agree upon. This may include time expectations and a general "work" agreement.

CASE STUDY: Mentorship

A research student commenced his PhD in an internationally recognized research group at a highly-regarded university. The group had about 20 members, including several postdocs, PhD students, Master students, and technicians. He met his group leader for about half an hour in his first week, and was assigned to his desk to design his research project with a one-page hand-written outline provided by his group leader/mentor.

Although the group had meetings every one or two weeks, the student was not working directly with anyone else in the group, and his attempts at establishing his experimental work did not take off. His mentor tended to wave him off, indicating not to give up and to keep trying, but provided little direction regarding whether his project outline was any good, whether it seemed like a viable prospect, or whether his project timeline was realistic.

The student ultimately quit his studies to move elsewhere for an opportunity he was offered in industry, which represented more money and a stable, permanent job that allowed him to have a more balanced life outside of work. He genuinely wishes to pursue a PhD during his career, because he has deep interest in the field.

This case study demonstrates how poor mentorship and guidance results in the loss of motivated and talented people. It is worth considering the potential outcomes to your approaches to leading and managing junior researchers under your guidance.

Table 3.1: One page mentoring checklist

A mentoring checklist like this should be completed by the new student/staff member and the group leader at the beginning of any new contract. Both should keep a copy for future reference.

Mentoring agreement checklist	
Duration of mentoring arrangement	
Purpose of mentoring arrangement.	
How often will we meet?	
What style will those meetings be? (Formal presentation, informal chat, coffee break; meeting duration, at least once every X weeks.)	
The role of the mentor is:	
The role of the mentee is:	
Our meetings will be confidential/not confidential.	
The mentor will provide honest and constructive advice regarding the mentee's work and progress.	AGREE or DISAGREE
The mentee will acknowledge the advice and take it into consideration in their research pursuits when actively pursuing their investigation.	AGREE or DISAGREE
The mentor agrees to provide feedback regarding written materials being published during the period of the mentoring arrangement.	AGREE or DISAGREE
The mentor agrees to promote progression of the mentee's work, wherever and whenever possible.	AGREE or DISAGREE
Authorships arising from the work will include the mentor as final author and mentee as first author, unless otherwise agreed upon.	AGREE or DISAGREE — CLARIFY
Training will be provided regarding: equipment, experimental approaches, project management, academic writing, conference presentation, other … (may relate to departmental requirements).	Specify training here.
Signatures: (Date and printed name)	

Dealing with Different Personalities in the Group

Everyone will encounter personal challenges in their pursuit of academic project excellence. Inflated egos and competitiveness are inherent in this professional sector, although they are not characteristic of all researchers. You certainly require a lot of internal drive and motivation to get past the myriad of stumbling blocks that you have to overcome. How you deal with them may simply come down to personality and experience. Some typical challenges often confronted by group researchers who are focused on achieving their project goals within the larger research group are discussed here.

Project direction conflict

CASE STUDY: Project direction conflict

Two of your group members have a difference of opinion about the direction a project that both are working on should take. This might be specific to the experimental direction for the next phase of the work, choice of collaborator, or even how the last of a limited reagent-stock is used. One of the group members has a dominant personality and you regard them highly, whilst the other feels that they lack much of a say in the direction of their own work. How do you address the issue?

Dominant personalities often draw the attention of group leaders for a variety of reasons. The only approach you can take to deal with a situation as outlined in this case study is to request a solid argument through logic, reason, and science. If your group member has a clear idea of why they wish to take a particular direction with their research, request that they formalize it. Suggest they create a strategy report that outlines the goals for the next stage of the work. Request evidence for their reasoning in terms of background and preliminary data, literature references, and a clear plan of experiments that will be pursued.

Encourage the team to treat the next phase of the work like a mini-project, and note down all of the expected benefits and challenges that will be faced. A solid scientific plan that is backed up with this type of clarity, a clearly well-thought out direction can ensue. Your team members may get a "feeling" about a particular direction, but the first and best way to back it up is to do their homework and investigate further before wasting time and money on a direction that may have been tried before, or which is likely to fail but the time was not taken to correctly research it before beginning work.

> ## WELL-REASONED PROJECT PLANS TRUMP
> ## PERSONAL OPINION

Well-reviewed literature can oftentimes lead to independent review publications. The value of taking time to review published work and data should never be underestimated.

Dealing with divas

> **CASE STUDY: Dealing with divas**
> A PhD student's project relied on the services provided by a departmental technician. This service involved the maintenance of a key piece of equipment that was essential to the progress of the research. For some reason, this technician decided that they disliked the student. This technician was important and central to the work, but they were also a diva. How should the student have worked around this to ensure the work's progress?

It is challenging to confront friction with people that hold sway over the progress of your research, as outlined in the case study above. When the group leader allows this type of individual to hold a position of power in the group, it can have disastrous effects. Similarly, your team has to acknowledge that someone who is so experienced with a particular technique or piece of equipment has to be extremely knowledgeable about it, which demands a certain element of respect from you.

This situation emphasizes how important it is to respect all members of the group and institution as a whole. This technician is likely happy in their career choice, and likely not going to go higher unless they make major changes to their career path. One way they can demand respect is to embrace any influence they have in the group. So rule number one: encourage your team to be respectful to this person. If your group members require their service, have them arrange a meeting to talk to them about it, without fear of asking for assistance. It is probable that many researchers ignore this person, as if they themselves are a piece of machinery, which must feel awful as a skilled professional. You have a lot to learn from everyone you work with, and each person has a background skillset that you cannot possibly measure.

THREE KEY APPROACHES
Advice to your group members when dealing with divas

1: Treat this person well and be humble. You might be doing a PhD, but this person may have many years more experience than you in this line of work.

2: Bring out the solid science to back you up if it seems like this person simply looks down their nose at you and is not at all willing to help. As in the recent case study on project direction conflict, nobody can argue against a solid scientific argument. Do your homework and write up a report to present in the lab meeting to explain exactly what needs to be done. Present your solid case to your group leader to argue the need to have this work done with the technician's assistance.

<u>3: Use your professional network.</u> If this technician works directly for your group leader, you need to convince your group leader that the work must be done and that they can instruct the technician to help you. Then they must do so. If they represent a departmental or institutional resource, then your group leader can act as your advocate.

Most critical in all of this is that you **maintain your professionalism**. This is a trait that is severely lacking in many aspects of academic research; nonetheless, when you show it, people remember and respect it. Do not let anger, annoyance, or nastiness creep into your workplace. Stay above it: be a professional.

MAINTAIN YOUR PROFESSIONALISM

Encouraging Positive Connectivity Within the Group

The style of management taken by the group leader can be the biggest influence on positive collaboration and networking within the group. Different group management styles are outlined in chapter 2. Beyond the group structure, various activities can promote collaboration and teamwork further, depending on the group dynamic and overall focal interests of group members.

1: Retreats

Group retreats can provide an outstanding opportunity for team members to build a solid working relationship. Despite cost, group members may look forward to these events as a highlight of their professional year. Retreats are great for groups that have enough money to fund them. Through a combination of team-building activities in the outdoors and daily discussions about the work, not to mention dinners as a group, group members can build strong relationships with each other, making a better working environment when you are back on the ground in your regular workplace.

2: Workshops

Workshops run over days or a week can be very helpful for a group to participate in developing their overall skillsets in their research area. In particular, when workshops require paired or group exercises, they provide an outlet for group members to bond in an arena that is relevant to, but independent of, the regular day-to-day activities of the research setting.

3: Journal clubs

Journal clubs are an excellent way to track recent advances in the field, practice critical analysis and presentation skills, and discuss how recent findings in the literature can impact the research projects that are currently underway or which will take place in the future.

4: Troubleshooting sessions

When any project reaches a stumbling block, it is worth scheduling a troubleshooting session to employ the critical analysis skills of the whole group to try to identify approaches that can be taken to confront the challenge. This provides a group-thinking mentality, where each participant supports other team members, and it opens up potential opportunities for within-group collaboration.

5: Regular group meetings

Weekly meetings are recommended in an active and productive research group. These meetings allow everyone to discuss their current work and address issues that may arise in the group week-to-week. Such discussions might include issues that need to be addressed in the research environment, introductions of new group members, or announcements of important events or findings.

6: Social celebrations

The benefit of celebrating birthdays, published papers, graduations, births, marriages, and other big life events for members of the group should not be overlooked in the social framework of any group. These are the things that make us human and connected.

7: Simple routines

Something as simple as having a regular daily or weekly lunchbreak together can impact the connectivity of the group. Morning coffee, Friday after-work drinks, or afternoon tea breaks can create a relaxed informal and social relationship between group members.

CASE STUDY: Morning coffee

During the lifetime of their tenure in a research group, a student evolved from being a junior technician to postdoctoral fellow. During this time, several senior research fellows were established in the group, all of which encouraged a regular morning tea coffee break over which everyone bonded.

In the beginning, this constituted a drip coffee machine in the group's tea room. By the end of their time in the group many years later, a professional coffee cart had established itself in a nearby courtyard and the entire group would sit on the steps in the sun outside the department at 10:30 every morning for their coffee break. This proved a time to talk about work, weekends, and life in general and formed a solid tradition for the entire team.

Promoting Group Collaboration and Cohesion

Positive collaboration is promoted within a group by taking away the intra-group competitiveness and encouraging all group members to support each other's success. Groups that do not collaborate tend not to take steps to encourage people to communicate and help each other openly. Communication with the group leader and between group members is key to working together cohesively. Similarly, as in any successful endeavor, there needs to be clarity regarding project boundaries and goals.

1: Regular touch-points/meetings

These can ensure all researchers are aware of the work each person is doing and allows them to openly communicate their basic weekly plans. This truly helps to ensure that nobody is out of the loop and to provide a basic working framework. Reinforcement of the fact that you will be each other's network in the future helps group members to appreciate the importance of working well together. Even if the group leader adopts a very positive group structure, cohesiveness can be lost in other areas (see the case study below).

2: Effective project management

Complete transparency and **communication** regarding individual responsibilities are key to maintaining collaborative efforts within the group. It is also crucial to promptly recognize individual accomplishments, contributions, or progress. Where projects are poorly managed, and when it remains unclear who will be recognized for the work until after it is completed, there is a high likelihood of friction between those that contributed to the work.

3: Open discussion about clear and individual goals

Discussing goals regarding the work that each group member is pursing, and their ambitions for the future, is essential. This ensures that each member works cohesively within the team.

CASE STUDY: Group collaboration

A research group is very well structured and is operating very effectively according to the team model (see chapter 2). Three new PhD students are due to join the team in the next academic year. The students are of different ages, but have a comparable academic background. The group leader decides to provide an extra financial top-up scholarship to the students as an incentive. The oldest, who has worked in another sector before commencing tertiary studies, receives the highest award. The middle student, who gained experience as a technician after their Master degree and prior to entering the PhD program, receives an intermediate award. The youngest, who has come directly from a Master program with no education breaks, receives the lowest award. These stipends and their value are known by the whole group and department. What is the outcome?

In this case, the students have been set up in a **competitive model**, where each is required to prove their worth against the others. The student with the highest stipend is likely to have a feeling of superiority over the others — a feeling that would be validated by their higher stipend. The group leader has taken an ageist approach to differentiate the highest-paid student from the others, who have an equal background with regards to academic merit. The lower paid students are likely to feel anger and discontent regarding the disparity in payment for an equal or potentially better job done, and are unlikely to want to help the higher-paid student in their work. Given that these students are set to work in the same group over a number of years, albeit on separate projects, this type of incentive disparity, which is not merit-based, is a set-up for disaster regarding group cohesion.

Underperformance from Staff

A certain amount of freedom is expected in most academic institutions, but if a project or group is managed well, underperformance should be effectively dealt with long before it becomes a major issue that sets back the project. If you have staff that consistently underperform despite the communication of clear professional expectations, the official method of handling them relies to a large extent on the departmental or institutional code of conduct. However, certain **checks and balances** should be in place long before this type of issue reaches official channels.

If you successfully interview, induct to the group, and communicate clearly with recruits regarding expectations and mentorship — in addition to maintaining regular meetings and involvement within the group — then underperformance should not be an issue. When things slide and leaders fail to lead effectively, staff have a tendency to lose motivation and become complacent and distracted, as if they have no goal to reach for and no outcome pending. **Promoting commitment from your staff** comes from promoting enthusiasm about the work, and from creating a positive network that encourages professional interest and direction in the work. Sadly, few group leaders take this responsibility to heart, and that is when you start to lose your staff's interest, and then underperformance becomes more common.

PROFESSIONAL GOALS PROMOTE INTRINSIC MOTIVATION

Job security is another issue that's frequently confronted in academic research. Most non-tenure staff are employed on short-term or several-year contracts. Few participants actually acquire job security at the technician or post-doctoral level. Given the competitive nature of research, it is important now, more than ever, that you support your staff and their development, and help them reach their professional goals. Key to that is identifying their professional goals (see "mentoring checklist" in this chapter). If the right approaches are taken to managing staff, including good communication, identifying the problems that staff are experiencing and which are causing their underperformance provides an avenue to reflect on the reasons behind their underperformance.

Academic leaders should follow a cycle of **planning, discussion, monitoring, and review** to stay on top of professional direction. This can be difficult to follow, especially when personal association is involved.

CASE STUDY: Underperformance

An established group leader hired a new postdoc based on recommendations from his colleagues. The postdoc had just defended their thesis and previously worked on his own small research project. The group leader held high confidence in the new postdoc and appointed him to coordinate a several years long large multi-center collaborative project. The group leader assumed the postdoc was doing a good job, and afforded him independence in coordinating the project. After a year, complaints are heard from the other collaborators: deliverables and deadlines are not met, crucial agreements remain unsigned, and the large project is in difficulties. The group leader and the unit/institute gain a bad reputation as lousy collaborators.

In the above case, what could the group leader have done differently? How could they have addressed the problem at the outset and reduced the escalation of project-related issues? What can they do in retrospect to correct the issues that have arisen?

Many group leaders simply hire staff and outline their expectations, leaving the staff member to handle it. Of course, this approach *can* work, depending on the person, but it is absolutely worth your while as a group or project leader to take some time to sit down and establish a full and open dialogue regarding professional expectations with all of your staff and students.

<u>Dealing with discontent</u>

Groups can sometimes reach a point where there is a general feeling of discontent amongst its members, often due to the poor approaches taken by the group leader. If you fail to manage the group well, proactively, and effectively, it is not unusual for the group to begin to rebel against the poor leadership. This type of discontent can equal the type of behavior that you might encounter in a school playground. The only way to deal with this is to **reliably inspire effective project and group management,** focused on professionalism and the work at hand.

CASE STUDY: Dealing with discontent

A group leader based in a prestigious university has lost interest in managing their group day to day, but instead has taken to employing talented people on high-paying fellowships and scholarships to work in his very highly-regarded research area. The acquisition of funding demonstrates to the faculty that the group leader is adding value. Nonetheless, once recruited, the junior researchers receive zero guidance and end up floating in their projects. They receive only criticism from the group leader, who is generally very self-focused and not at all supportive.

Eventually, even the most motivated of these researchers loses all of their motivation, and they focus on doing the bare minimum need to survive this toxic work space, focusing much of their attention on how they will leave the role. Whenever the group meets socially, they complain extensively about how ineffective the group leader is at managing the group.

In this case study, the group leader has lost sight of the fact that they have to maintain the productivity of their group through effective management, instead focusing more on departmental politics and securing regular funding to show that they are bringing money into the department.

This type of spin-doctoring is rife in academic circles. This ethically represents not only disrespect to people that have chosen the group leader as a professional mentor but also a tremendous waste of funding on research that ends up routinely going nowhere. Many junior researchers who enter a group like this end up leaving academic research, disillusioned and disappointed that they have wasted so much of their life on this "idealistic pursuit."

END OF CHAPTER 3 SUMMARY
Human Capital

In this chapter we have discussed issues relating to human capital — the people who constitute the engine that drives your research forward. It is essential to proactively develop and manage your research group. The main foci of this chapter included:

- Ensuring your team is professionally satisfied and supported is critical for the retention of skilled people.

- There are various places you might source staff and students, as well as funding to pay them.

- We discussed the role and value of each group member, and how new staff and students should be inducted to the group.

- Good mentorship strengthens the performance of your group members.

- A professional approach and attitude is essential when dealing with challenging personalities in the group.

- Various approaches can promote positive collaboration within the group.

- Effective management of, and communication with, your group is absolutely essential to minimize issues related to underperformance and discontent.

REFLECT

1: How have you sourced your staff in the past, and have you been satisfied with their performance?

2: How has the structure of your research group impacted your time and productivity? How do you feel you could improve on this?

3: What personality issues have you seen in your academic experience? How were they navigated? Could it have been handled better?

4: What approaches have you seen or used to promote within-group collaboration and support?

5: How would you best deal with a staff member or student who is underperforming in your group? Do you proactively communicate your expectations to your team, individually and in working groups?

EXERCISE 3.1: STAFF AND RECRUITMENT

Make a list of all of the staff and students you have employed or supervised during your research leadership career. You can download an example and this blank spreadsheet from the Practical Academic website (www.practicalacademic.com). Consider how you sourced these staff, and how successful each has been in your group. What defined their success?

EMPLOYMENT HISTORY RECORD: Group name				
Group focus				
START DATE	END DATE	NAME	CONTRACT TYPE	WORK FOCUS

EXERCISE 3.2: INDIVIDUAL STRENGTHS AND GOALS

Write out a list of all of your current staff and students and note the criteria in the following table for each staff member. List the person's professional and personal strengths related to the work that they are pursuing in your group. Also list the person's professional and personal goals, as you understand them, which they hope to achieve in the timeframe of their role in your group or over the next few years. Note for each person, as their group leader, how you help them to maximize on their strengths, address their weaknesses, and reach their personal and professional goals.

PERSONAL AND PROFESSIONAL DEVELOPMENT	
STAFF MEMBER NAME	
PROFESSIONAL STRENGTHS	
PERSONAL STRENGTHS	
PROFESSIONAL GOALS	
PERSONAL GOALS	

Chapter 3 — Downloadable Materials
Download from www.practicalacademic.com

- Table 3.1: "Mentoring checklist" — Word file containing a template.

- Exercise 3.1: "Employment history record" — Excel file containing a blank template, an example page, and an extended template.
- Exercise 3.2: "Personal and professional development (individual strengths and goals)" – Excel file containing a blank template and an example of the completed exercise.

REFERENCES – Chapter 3

Bozeman, B.; Feeney, M. K. (October 2007). "Toward a useful theory of mentoring: A conceptual analysis and critique". *Administration & Society*, **39**(6): 719–739.doi:10.1177/0095399707304119.

CHAPTER 4

Group Organization and Logistics

A central aspect to establishing a functional and productive research group is the level of institutional support that provides infrastructure to pursue the research in question. It is widely acknowledged that research and office space, and access to water, electricity, and basic amenities are key considerations when establishing any newly-arrived experimental-focused group in an institution. Nonetheless, several other factors can also significantly contribute to progress, and they will be discussed here.

Budget Management Systems

Academics are not renowned for their budget management skills. After at least six years' tertiary training, followed by postdoctoral investigations, the majority of your career has been gauged by your ability to do good research, not necessarily on your ability to manage a lab. Once you enter an academic institution as a group leader with your own budgets and people to manage, in addition to all of the other responsibilities that you are saddled with, your time in the research environment tends to dwindle; you become a glorified middle manager, albeit usually with no management training.

When it comes to budgets, a lot of us are in the dark and reliant on our own personal histories regarding how to manage finances. Your institution may have a **centralized budgeting system**, with an accountant who balances the books and is available for consultation. This resource can be a huge benefit; however, this support should not negate the individual responsibility of keeping track of finances. **Clarity in budgeting** between the funding bodies, the institutional players, and any groups involved is key to a successful research undertaking. When you consider that a vast majority of academic research is funded by government-derived money, which essentially is funded by the taxpayers, transparency is what we all need and want. You have a responsibility to spend and manage your research funding conscientiously. Think of your research as a business owned by your funding investors, and consider that you are responsible for producing certain output (journal articles, tools, intellectual property, and so on) within a certain budget, to maximize your "business" success.

Managing the Group Budget

Establishing the budgeting tools for your institution is really within the remit of the departmental managers. However, as an independent group leader, you should be keeping track of your own spending. Researchers may follow a range of approaches to ensure that their group stays funded, so that projects can continue.

Delayed funding strategy

Some groups acquire new grant money for work that is already half completed. They might acquire this through short exploratory projects or as an adjunct to other work. Using this approach to manage the funding of research pursuits, a group can actually free up funds to allocate to other exploratory projects, viewing the group money as being stored in a budget repository to benefit all group members. This requires expenditure to be tracked for a given project to ensure that project reports are well structured, and that individual project spending is kept clear on itemized accounts for funding bodies.

Direct funding strategy

When you follow the standard approach in which you commence your research after obtaining your funding, there is tremendous pressure to achieve deliverables within the timeframe outlined in your original grant application. This approach is more common with junior researchers and small research groups that lack a solid, established research program. It presents a greater risk of incompletion, given the likelihood of project delays that may extend beyond contingency time allowances included in the project plan. Larger groups tend to have the capacity to incorporate other funding strategies to confront impediments to project progress.

"Saving for a rainy day" strategy

This strategy involves requesting a strong budget on a project, and then working effectively under budget to achieve the goals set out. This creates an overall positive balance in the grant account that can be utilized for other exploratory project pursuits or to extend the funded project further than outlined, potentially creating preliminary data for future grant applications. This approach can represent an opportunity for researchers to extend their research investigations over time, and thus extend their overall research foci. Ways to approach this might include seeking collaborations to achieve key components of research investigations, or acquiring scholarships for research students on the project. Such scholarships would include research funds that can be funneled into the project funding pot. This provides the bonus of acquiring extra project staff without dipping further into the budget.

Contingency budgets / slush money / soft money

Both the delayed funding strategy and the "saving for a rainy day" strategy tend to create contingency budgets that can be used to cover extra costs that inevitably come up on projects. Such costs can change quickly for any reason (e.g. novel technology available at a cost, extra costs to prove an unknown variable, requirement for extra reagents, and so on). Most larger labs will have a contingency account that the group leader keeps on hand for useful expenses, such as laboratory tools, reagents, travel, or extra staffing on intensive projects.

THE FUNDING BODY DESERVES RESPONSIBLE MONEY MANAGEMENT

Issues in Budget Management

A number of issues can arise, thus groups should not depend predominantly on central accounting systems. Various approaches can be taken to work around these issues.

Ordering

Keeping an Excel **ordering spreadsheet** that is backed up to a cloud drive or core hard drive each week, or even kept in a printout folder once each page is filled, is a proactive way to keep track of group spending. Other specific software packages for keeping tabs on grant spending are available commercially from various sources, but a simple Microsoft Excel spreadsheet does normally suffice. This type of tracking system can be regarded as a lab-managed tool to keep track of what you are expecting to be delivered from orders, details of product numbers, amounts spent, grants used for payments, and any other relevant information (see table 4.1). This is also a good way to keep note of any special offers that your group has negotiated with suppliers, and things can be marked as delivered with a note of their storage location by whoever receives the items. The person who receives the items should be responsible for notifying the person listed as having ordered the product. In the spreadsheet, the background cell can be colored gray once the product has arrived, for ease of reference about deliveries still pending. This type of spreadsheet can be referred to each month to compare the departmental budgeting records for your grants, to ensure that there are no budgeting oversights. This keeps everyone accountable and provides a backup to compensate human error.

Table 4.1: Ordering spreadsheet

An example of an ordering spreadsheet in Microsoft Excel, used to keep track of lab expenditure and orders. An extra column may be included to record information such as special storage conditions, reference to notes in a lab book, safety issues, or specific work that the item is ordered for. A digital copy of this example is available on the website (www.practicalacademic.com).

ORDERING MANAGEMENT SYSTEM: Xavier Marques Group									
ORDER DATE	ORDER #	ACCOUNT	ITEM	Cost	SUPPLIER & CAT#	ORDERER INITIALS	RECEIVED DATE	RECEIVED BY INITIALS	STORAGE LOCATION
14/07/2016	2016-98	2016/A1	Restriction Enzyme - BstXI (5000U)	$262.00	NEB - R0113L	NM	27/07/2016	TM	-20 main lab
14/07/2016	2016-99	2015/C3	NucleoSpin Plasmid EasyPure (250 preps)	$493.00	Scientifix - 740727.250	AP	19/07/2016	TM	AP bench
14/07/2016	2016-99	2015/C3	NucleoBond Xtra Midi (100 preps)	$1,317.00	Scientifix - 740410-100	AP	19/07/2016	TM	AP bench
15/07/2016	2016-100	2016/A1	RPMI1640-L-glut 500ml x 14	$280.00	Sigma Aldrich R0883-500ml	TM	20/07/2016	RL	cold room top shelf
15/07/2016	2016-100	2016/A1	DMEM - 500ml x 14	$217.00	Sigma Aldrich D5546-500ml	TM	20/07/2016	RL	cold room top shelf
15/07/2016	2016-100	2016/A1	FBS - Sterile Filtered - 1L	$554.00	Sigma Aldrich D5546-500ml	TM	20/07/2016	RL	TC Freezer
19/07/2016	2016-101	2014/A2	Dissecting Forceps- 22cm	$189.32	Surgical Instr 17.21.80	KKT	27/07/2016	AP	KKT bench
19/07/2016	2016-101	2014/A2	Adson delicate forceps - 15cm	$70.23	Surgical Instr 06.21.15	KKT	27/07/2016	AP	KKT bench
19/07/2016	2016-101	2014/A2	Blank dressing forceps - 13cm	$33.04	Surgical Instr 06.04.33	KKT	27/07/2016	AP	KKT bench
19/07/2016	2016-101	2014/A2	Micro forceps ultra fine - 14cm	$248.39	Surgical Instr 07.63.30	KKT	27/07/2016	AP	KKT bench

An example of an ordering spreadsheet in Microsoft Excel, used to keep track of lab expenditure and orders. An extra column may be included to record information such as special storage conditions, reference to notes in a lab book, safety issues, or specific work that the item is ordered for. A digital copy of this example is available on the website (www.practicalacademic.com).

Tracking expenditure

CASE STUDY: Tracking expenditure

A well-established research group has approximately fifteen people working across several lines of investigation. The group operates a modest budget. It has some high-cost disposable items maintained as resources in the central stocks, including enzymes, reagents, radioactive labels, extraction kits, and polypropylene tubes and materials. A technician is tasked with maintaining the central stocks and keeps a central Excel spreadsheet to track every order made for the group. Individual users also add their special-order items to the spreadsheet, so that they can be tracked for delivery and location when they arrive. This is coupled with a paper-system linking the signed-for delivery sheets, which are placed into a folder for delivered materials by the person receiving the item. In this way, the tracking of ordered materials is accessible and traceable from the moment an item is requested via the online ordering system. This proves particularly useful when items go missing.

Schedule time each month to go over the expenditures for your group to double-check that you are on track. Rationalize the work done against that which you have budgeted for, and forecast potential future expenditures for the next month. If your group or project tends to use the same basic expenses each month, you can plan your budget for the whole year in a process known as **"calendarization."** This facilitates planning for exceptional expenses, such as a required machine or other large equipment purchase.

Monthly budget checks represent a good time to also schedule meetings with project groups, where you go over their expenditures on a given project to discuss the limitations or needs for the next month. Including all staff that use resources in your group in the budgeting discussions only helps to encourage joint responsibility, and thus a more collective conscience for spending on-budget. This is also a good time to reflect on what kind of funds will be needed for future work that you might spin out into more funding applications that you can apply for; in particular, you might apply for fellowships and scholarships with your staff or students, which may include extra funds for research expenses that can complement the grant money that you have already been awarded. As long as the fellowship is focused on salary, and is not purely grant money, this is not necessarily regarded as double dipping on one project. Instead it is regarded as possibly expanding project potential by freeing up more funds to employ more people, or to access more resources, to promote project progression.

Fellowships and scholarships

Fellowships and scholarships that are so hard won require solid management in order to ensure that you maximize potential. This begins at the application stage, where the applicant researches and acquires funding with the best potential conditions. It pays to read the fine print regarding what the associated reimbursements are. Fellows and scholars may be entitled to, or given, significant extra benefits in excess of the basic stipend that they are awarded. These benefits may include, but are not limited to, IT budget, research expenditure, travel money for conferences, and relocation allowance.

When a fellow or scholar receives an award, it is wise to insist that they take responsibility to manage their own stipend, because — much like the work that they will deliver in their project — they have the most vested interest in using the award to its full advantage. This represents an opportunity for the fellow to establish professional skills in budget management, and will likely aid their appreciation for experimental costs, and thus for appropriate planning. As a group leader, you should keep track of your fellow's spending, but **promote self-efficacy** through which they direct their own money management under your guidance.

CASE STUDY: Poor money management

A postdoc arrived in a new institution to commence their work, and they were funded by a fellowship. The fellowship came with some research money, which the group leader insisted on managing. After nearly two years, the postdoc visited the institutional accountant to arrange some money to attend a meeting, and it became clear that the research funds had not been recorded correctly for two years. Now the postdoc had three days to spend the funds, which were substantial, or the money would return to the funding body.

Databases

Similar to the spreadsheet described for ordering, Microsoft Excel spreadsheets can facilitate the management of **databases of reagents, stocks, and other materials** that are present in the lab. Some departments or institutions keep an **institutional record of equipment** that is updated annually in time for major grant applications of all of the equipment that the department has available for use. In such cases, it is used as a support document to append to grant applications.

Each group should also keep track of what is available in their research space. This might be a register of chemicals, a register of animals, a register of cell or DNA stocks, a register of antibodies, or a register of samples or reagents. Only good management will yield good research outcomes, and thus good databases are essential for effective research pursuits.

GOOD DATABASES ARE ESSENTIAL FOR EFFECTIVE RESEARCH PURSUITS

Safety

Ensuring your group is operating in a safe environment should be centrally regulated by the institution that you are affiliated with.

However, it is also your responsibility as a group leader to ensure that your group is working in a safe space. This may relate to laboratory investigations or field work. **Safety first, always**.

If you have fellows and students entering your research space, encourage them to question safety in the workplace, which often depends on the overall culture of the institution. Safety induction sessions should be run by departments at the commencement of any new appointment, and project/lab-specific inductions should include any safety issues that will be encountered by the new appointee. Nonetheless, given that a large chunk of academic experimental research employs cutting-edge techniques and equipment, there is often a lag between the method you employ and the established safety regulations. In this case, the onus is on the person responsible for the work. In the academic research group, this is typically the group leader responsible for hosting the work in their research group.

CASE STUDY: Safety

A junior researcher was working with cells in a sterile flow cabinet, testing cell behavior in response to different treatment types. The researcher was struggling with their project after several weeks, because all of the assays failed. During their biweekly meeting, the supervisor assumed the student must be making errors, and thus requested a larger single experiment carrying all controls. The student then spent three hours using the sterile flow cabinet to set up this large test experiment — a much longer duration than usual. Immediately following this time, the student was rushed to hospital, unable to see.

When setting up the flow cabinet for the day, the student was not switching off the UV light used to sterilize the working surface. All the cells had been dying due to UV stress during the experimental setup. The student thus sustained burns to their eyes, burning their corneas. Were safety guidelines adequately provided or followed in this case?

THE GROUP LEADER MUST ENSURE SAFE WORK PRACTICE

<u>Keeping track of safe work practices</u>
The above case study demonstrates the importance of establishing and tracking safe work practice in the research environment. In order to best support safety in your group, standard operating procedures (SOPs) should be developed to address local laboratory operations and practices. SOPs should be stored in print form in an accessible location for all people working in the research space. SOPs should include safe work practices for all routine protocols employed in the lab. Safety sheets should be accessible for reference.

SOPs are standard in industrial research settings, but are not yet standard practice in the academic setting. Nonetheless, your institutional safety officer should be able to provide you with guidelines regarding how best to standardize your group's SOP database.

Support / Infrastructure

Support staff and facilities are another factor that can contribute to the productivity of your lab and projects. It is well known that institutions providing a lot of facilities and support services attract "better" candidates. This relates primarily to group leaders with established grants, those that publish well, and those who have great professional standing amongst their peers. This is certainly something to consider when students and staff are discerning where they want to work, because better facilities enhance the likelihood of productivity. Of course, even if you work in the best facilities in the world, if this fails to combine with a sound research plan and good communication/academic connection between research participants, the work will suffer.

Even in smaller institutes, research groups that are established well can still utilize solid management to outperform those with more resources at their fingertips.

CASE STUDY: Support infrastructure

A small group began to grow, and the group leader decided to help the productivity of his researchers by employing a cleaning lady to wash dirty glassware and autoclave sterile materials three mornings a week. This represented a small cost outlay, but significantly impacted the group's collective productivity by omitting the need to each individually prepare their own materials. The cleaning lady also presented a positive dynamic to the group, promoting a pleasant social atmosphere.

The same group later moved to a more competitive, well-funded, shiny new institute, and this institute had a centralized cleaning arrangement where glassware was placed in tubs and sent to the kitchens to be washed and autoclaved. Sterile packages of materials were available around the clock in a large cupboard for all. But the pay-off was not necessarily worth it for the group. The productivity remained the same, or even declined, because there was a lot more pressure to compete against the other researchers in the institution and there were many more meetings to attend. The group was dispersed into cubicle desks where there was less opportunity for networking within the group and thus lost much of its social identity.

This case study demonstrates the definite advantage of establishing support staff or infrastructure, and it also emphasizes that it is only one of many concerns when you are managing a research group.

Support staff can be key to productivity, whether they are hired in-house or within the institution in which you are established. Expert workers all play a key role in the effective functioning of your research.

This may include technicians, microinjectionists, tissue culture managers, equipment managers, FACS operators, finances/ordering staff, receptionists, lab managers, cleaning staff, departmental managers, and store staff. These people must be respected for their contribution.

> ## HIRING SUPPORT STAFF CAN BE BENEFICIAL TO PRODUCTIVITY

Shared Responsibility for Tasks

When you consider that your PhD students and postdoctoral fellows are essentially training to be able to run research projects, it is important that you train them to be able to manage all aspects of the research group. This includes budgeting/funding management, databases, stocks/reagents, ordering, and safety. These aspects of research group and departmental organization should not represent some mysterious component of research life. Everyone in your group should be aware of how the research operates in your group and department.

One way to maximize the capacity of your team whilst minimizing distraction away from their research is to develop a duty roster, assuming your group is large enough to carry it. Regular weekly or monthly tasks may be allocated to a given person. Some examples of duties are noted here; these form a representation of classic roster duties.

Typical roster duties

Some typical duties that may be included to a duty roster for your group are listed here. These may be related to the example duty roster shown in table 4.2.

- Replenishing stocks of water, phosphate buffered saline, or other standard reagents.
- Managing the ordering system.

- Emptying large laboratory trash (if applicable).
- Coordinating a monthly meeting or social event.
- Repairing and maintaining shared equipment and related consumables.
- Check lab first aid and chemical spill kits, eyewashes, safety showers and extinguishers.
- Update records relating to safety, standard operating procedures, and group databases to ensure paper and digital records correspond.
- Clean and maintain research spaces.
- Clean and maintain research equipment, fridges, freezers, cold rooms, or similar.
- Ensure stocks of chemicals or other materials are in order.

CASE STUDY: Rostering duties

A young group leader develops a duty roster for his team of five. The group is highly experimental, thus generates a lot of waste, and utilizes a lot of prepared reagents day-to-day. Everyone gets along for the most part, despite each group member working individually.

An international postdoc joins the group after a year of an established roster system, which has been working well. The postdoc is poised to commence their own research group after leaving this institution. The postdoc deems their time to be more valuable than that of the remaining group members, who are mostly postgraduate students, and thus completely ignores the established duty roster system.

The junior group members get angry with the postdoc for not helping. The postdoc never contributes to tidying, cleaning, or replenishing reagents. At times, he uses communal resources and does not replenish them. The group leader is petitioned by his angry group to address the issue directly with the postdoc.

Table 4.2: Example duties roster for research group

An example of a group roster in Microsoft Excel, used to allocate duties fairly among group members to keep the experimental space running smoothly. An extra column may be included for the assigned person to write their signature to acknowledge that the duty has been performed, and on which date this took place. A digital copy of this example is available on the website (www.practicalacademic.com).

Table 4.2: MONTHLY DUTIES ROSTER: Xavier Marques Group

Duty	JAN	FEB	MAR	APR	MAY	JUN	JUL	AUG	SEP	OCT	NOV	DEC
Replenish MQH2O and PBS stocks	NM	RD	FS	LS	AT	XM	KKY	TM	NM	RD	FS	LS
Change trash liners and replenish EtOH 70% stocks	TM	NM	RD	FS	LS	AT	XM	KKY	TM	NM	RD	FS
Clean out Fridges and Freezers to reduce ice build up and wipe down surfaces	KKY	TM	NM	RD	FS	LS	AT	XM	KKY	TM	NM	RD
Clean out Cold Room and Storage Cupboards	XM	KKY	TM	NM	RD	FS	LS	AT	XM	TM	TM	NM
Tidy up chemical stocks storage– alphabetical order	AT	XM	KKY	TM	NM	RD	FS	LS	AT	XM	KKY	TM
Weekly Ordering	LS	AT	XM	KKY	TM	NM	RD	FS	LS	AT	XM	KKY
Sync paper and digital databases	FS	LS	AT	XM	KKY	TM	NM	RD	FS	LS	AT	XM
Chair monthly journal club and lab presentation	RD	FS	LS	AT	XM	KKY	TM	NM	RD	FS	LS	AT

In-house Data Sharing and Communication

How does your group share information internally? Most have established email connections, but do you support shared drives, paper trails, collections of materials as resources, or digital organization systems? Effective data sharing and storage are key to well-organized and productive research outcomes, and also support solid research development.

Key digital management approaches

1: **Make an email group** specific to your team, so that all group-specific matters can be announced directly without having to type out each name. You can add or remove people as they travel through your group.

2: **Set up a shared drive** to hold all of the laboratory's digital resources, including digital databases, records, standard operating procedures, and information about resources.

3: **Use a calendar system** to schedule meetings and events. Regardless of the system that your institution employs, you should be able to establish a digital calendar system as a normal operations tool in the management of your group. Scheduling meetings, journal clubs, social events, and workshops via this means will simplify communications, ensuring that the staff and students working with you can easily track activities.

4: **Digital lab books** or experimental notes may be employed to record experimental data, just like a paper lab book. A number of different digital lab books have arisen as useful tools with the intention to replace paper lab books. Whether you pursue this or not is entirely your decision, but it is worth exploring, particularly if your work is very informatics-focused with a large amount of digital data. Digital lab books make it easier to replicate and store experimental records, but they can prove challenging for traditional lab-based and field-based researchers.

5: **A group webpage** represents a positive way to publicize the work going on in your group. It is also a lure for potential collaborators, who will be able to gauge the value of the work being performed. Group members benefit from a public platform explaining their current career activities, and potential future group members can gauge whether the group appeals to them. In addition to this, webpages may include login sections that can be used for digital collaborative purposes, carrying forum discussions or other methods of communication and data sharing.

6: **The backing-up of digital material** is critical to maintaining your records. Many institutions, faculties, or departments will host shared drives that automatically back-up to another location at the end of each day or week, whilst others provide little or no digital back-up support. You need to gauge the best options for your situation, and determine what actions you need to take to ensure that the data being generated by your group is secure and will not be lost after one hard-drive crash. It is critical to share this with your staff when they commence in your group.

END OF CHAPTER 4 SUMMARY
Group Organization and Logistics

In this chapter we have discussed issues relating to general organizational structure, and how it can promote productivity in the research group. All of the systems discussed in this chapter should be openly discussed with all group members when they join the group. An overview of how these processes are controlled in your group should be available for all group members to review and agree to. Several key take-home messages are delivered in this chapter.

- Proactive efforts should be directed toward managing the group's budget and establishing systems that are able to track expenditures and finances.

- Employing strategy in active budget utilization can enhance your productivity from the funding that you have won to advance your research.

- Databases help to organize the materials that you have in your research space, and the materials that you create as part of the research process.

- Safety is your responsibility as the group leader, and you must ensure that all group members are acting safely in the research space.

- Support infrastructure can enhance the productivity and appeal of your research group, but it must coincide with sound management practices.

- Establishing duty rosters can enhance the smooth operations, whilst encouraging your group to collaboratively learn and contribute to all aspects of research life.

- You should form an established practice for managing digital content that is generated and stored by your group.

REFLECT

1: How have you managed your budgets in your research pursuits?

2: Many group leaders retain an air of mystery around group budgets, so the researchers themselves rarely have any idea of the current funding. Reflect on your professional experiences and how open the group you were associated with was about budgets. Did it work well? How do you think it could have been done better?

3: How do you ensure that your group is following safe practices? How do they record their materials in centrally accessible databases? Does your group work together effectively and share duties?

EXERCISE 4.1: BUDGET STRATEGY

Make a list of the various funding sources from which you have derived funds during your career as a group leader, and gauge how those funds have translated to various outputs. Include fellowships, grants, and unusual sources of funding support.

Funding source	Outputs/deliverables

EXERCISE 4.2: DATABASES

Make a summary of all of the current databases that you have established in your group, noting who manages them and how effective they are. Does your group keep them up to date; do you ensure that everyone contributes? What might you implement a database for?

Database	Manager	Current usage

EXERCISE 4.3: DUTIES ROSTER

Download the duty roster template from the website (www.practicalacademic.com) and complete a roster for your current group. Consider all of the duties that you think it fair to include, and determine whether there are any yearly or twice-yearly clean-ups you may include into the group's annual schedule. Refer to table 4.2 for inspiration.

Chapter 4 — Downloadable Materials
Download from www.practicalacademic.com

- Table 4.1: "Ordering spreadsheet" — Excel file containing a blank template and an example page.
- Table 4.2: "Duties roster for research group" — Excel file containing a blank template and an example page.
- Exercise 4.1: "Budget Strategy" — Excel file containing a blank template and an example page.

- Exercise 4.2: "Databases" — Excel file containing a blank template and an example page.

CHAPTER 5

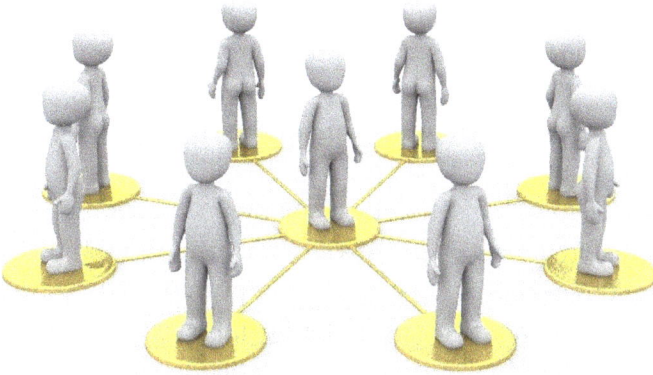

Networks and Professional Development

Similar to the business sector model, a key component of the academic research sector is the networking and mentorship that is integral to the advancement and management of research projects. This has already been discussed to some extent in chapter 3, but here we expand our discussion of professional networking and mentorship. While consideration must be given to discipline-specific concerns, some general approaches can be considered as tactical for professional networking in the academic research sector.

Networking is important in any professional endeavor. In both business and academia, networking aids the development of professional connections. In business, these connections might relate to professional opportunities or, at the operational level, sourcing better suppliers/supplies, deals on purchasing, or expanding the customer base. In academia, networking relates more to access to resources or knowledgeable collaborators that can provide more effective and productive research outputs.

Networking can be pursued on a number of levels, from junior researchers to senior faculty members. Both internal and external networking play an important role in the life of any business or

97

academic institution. Indeed, much insight can be applied from practices that are common in the private sector. Different networking approaches will be discussed here, taking into consideration the point of view of different roles in the research group.

Networking Outside of the Institution

<u>Conferences</u>

The principal form of networking for the vast majority of research academics is attending conferences, both national and international. Although conferences are outstanding opportunities to **gain insight** into the most recent advances in the field, they are also a wonderful chance to **meet and communicate** with people that you might want to work with, in some form, in the future. For more junior researchers, this might take the guise of wanting to go and work with another group, whilst more senior researchers may be seeking collaboration possibilities or resources that could advance their work. Established researchers in a field may look forward to the major conference in their discipline each year as a chance to catch up with old friends and colleagues; it also represents an opportunity to discuss their work with peers.

Good preparation is standard for attending conferences, given that attendance lists are typically made available prior to attending. You should also **bring a notebook** to keep all of your notes from the conference in one location, so that you can refer back to them.

Preparing for conference networking

Whilst attending sessions and talks can be useful, many researchers claim that the networking that takes place *outside* of the main sessions is the most important aspect of attending conferences. In order to identify with whom you would like to meet, it is useful to **start by reading the abstracts** available for the conference, in addition to identifying **who is attending** without presenting. Most conferences provide abstract books prior to commencement, which provide the opportunity to gauge the most interesting presentations that are relevant to you and your interests. You can build a schedule of attendance for each day of the conference, and include into this schedule any meetings that you might arrange in advance. The contact details for all participants should be included in the abstract booklet to facilitate this purpose.

It is worthwhile to **create your own agenda** for the conference to reduce confusion and keep on track, because you are likely to get distracted by different activities throughout the conference. An example agenda for day one of a four day conference is shown in **table 5.1**. It takes time to prepare this type of agenda, but it is worth your while to be organized given the limited time you will have, because it will maximize the value you gain from your attendance. Similarly, if you plan to report back on the conference to your group, you can refer back to your agenda and any notes that you take throughout the conference.

Meeting at a conference

> **CASE STUDY: Meeting at a conference**
>
> A new postdoctoral researcher has developed a tool that they would like to employ in combination with some resources already published by a group from another country. The postdoc has never met the professor that published the work, but sees that they are a keynote speaker at a conference being held nearby in a few months' time. The postdoc arranges to attend the conference and communicates with the professor by email, arranging a five minute meeting during the conference. This results in the postdoc pitching the collaboration successfully, resulting in funding spin-offs and several reciprocal collaborations spanning many years.

If you want to specifically meet somebody at a conference, it pays to arrange a meeting in advance. This may take place during poster sessions, during between-symposia coffee breaks, during one of the session intervals, or in the evenings between daytime conference schedules. You should generally prepare an agenda of concerns that you would like to broach with anyone that you meet in the conference setting, and bring with you any support information you might need. Many of these materials can be transported easily on a laptop, and you should have the materials ready to share by paper, or by attaching them to an email that you can send immediately before or after the meeting.

Materials you might bring to a meeting at a conference

PhD student/postdoctoral fellow seeking next host group:

- o guidelines for potential fellowships that you might apply for,
- o outlines for projects you'd be interested in pursuing,
- o copies of papers you have published for their reference,
- o a copy of your CV, and
- o references from previous mentors.

Researchers seeking collaboration/resources:

- o preliminary data outlining the relevant project to date,
- o project plan with projected timeline,
- o grant application relevant to the work, and
- o questions/queries about the potential collaboration you'd like to pursue.

Table 5.1: Agenda for conference (Based on Endo 2015 schedule)

DAY 1: AGENDA		
TIME	**EVENT**	**LOCATION**
8:00 AM – 9:15AM	L1 – Plenary	Hall G
9:30AM – 11:00AM	S4 – Symposium – IGF	Room 23
11:15AM – 11:30AM	PP02 – Poster Session – Endocrine Neoplasia	Room 6B
11:15AM – 11:45AM	OR10 – Award Lecture – RE Weitzman Inflammation/Metabolism	Room 20A
11:45AM – 1:00PM	OR10 – 10 min talks – Inflammation and Metabolism	Room 20A
1:00PM – 3:00PM	LUNCH Posters: LBT-(051-054),(056-063)	Hall D-F
2:00PM	*Meet A.Loppianen – Bring support data for collaboration discussion (2PM: ENDO café)*	ENDO Café
2:30PM – 3:15PM	MEIC1 – Editors Endo/MolEndo	Room 7
3:15PM – 4:15PM	L2 – Plenary	Hall G
4:30PM – 6:00PM	S16 – Symposium – Endocrine Cancers	Room 24
	ES11-23 – Basic session	Ballroom 20D
6:00PM – 7:00PM	Rest Break	
7:00PM – 8:00PM	Meet G.Beco for discussion re: project	ENDO Bar
8:00PM – 9:00PM	Dinner with Beco, Smart, and my group	Chillis Bar and Grill

https://endo.confex.com/endo/2015endo/webprogram/programs.html

Networking opportunities associated with conferences

If you are attending a very large international conference, you might discover some particularly interesting **satellite programs** or symposia targeted to certain topics. These can often also represent the annual meeting of an association, given that many members are likely to attend the larger conference. Similarly, **pre-conference events** or **post-conference events** (like practical workshops or training sessions), in addition to career forums, may also be advertised with the conference material, but they may have separate registrations.

Summer schools

Institutions and research organizations sometimes offer summer schools for postgraduate and postdoctoral researchers. These are particularly related to networking and skills acquisition. These provide an opportunity to develop more in-depth links with peers that are pursuing similar work to your own. Teaching on summer schools can also strengthen your network significantly.

Direct inquiry

CASE STUDY: **Making a direct inquiry**

After completing the development of a new test system as part of a Masters project, a first-year PhD student wanted to expand the battery of tests done on the system to consolidate its usefulness. The student contacted an international group leader who had developed the test that was already in use, and asked whether they were interested in allowing the student to set up the new test system in their group in exchange for access to the extensive sample collection available in the host group. This simple inquiry led to the successful receipt of international-collaboration travel grants, a multinational collaboration, and a number of publications comparing and contrasting the two test systems.

If you have identified someone who can provide you with a resource or opportunity that may be able to advance your work, you do not need to wait until the next conference to establish communication. Making a direct inquiry to other researchers has never been easier, given the simplicity of communication in the digital age. How you go about inquiring will depend to some extent on your position within your organization and what it is that you are seeking from the person to whom you are writing. Similarly, their standing in their organization, or in the field as a whole, may dictate the likelihood of their response. A PhD student from a small institution may find it challenging to receive feedback from a lab leader who directs an institution and is shortlisted for a major international prize, for example. That said, it never hurts to try.

Through people you know / referral

Group leaders usually have quite **established professional networks**, often having worked in their field for a number of years since commencing postgraduate studies. Therefore, students and postdocs can often acquire these networks themselves, or at least request resources from their supervisor's network.

Professional networks can, however, change and evolve as you move through different specialties that you may require over the expansion of your research interests. Thus, you will likely develop and expand your professional networks as you progress through your career. The advantages to meeting potential professional links through people with whom you already have an established working relationship cannot be overlooked. Similar to the recruitment of staff you may already have worked with, or who come highly recommended, networking with people who are well known by one of your trusted contacts can strengthen the faith and trust that you are likely to place in new professional collaborators.

<u>Organizational and digital networks</u>

An increasing number of organizational networks are being established internationally to promote connectivity between academic researchers. Organizational networks have grown dramatically in scope with the advent of the digital era, because a networking forum is often established in tandem with an organization's website. Networks may be promoted through **established online resources** such as LinkedIn, Twitter and, to some extent, Facebook.

With the growth of social media, virtual support networks for postgraduate and postdoctoral fellows have grown significantly. Twitter, Facebook, and independent websites host networks providing support, interaction, and advice regarding professional development and advancement in research pursuits. These can also be hosted by major institutions that develop virtual networks available to postgraduate students internationally. Such networks include journals, societies, university-led networks, and discipline-specific organizations. See appendix 1 for further information on virtual networks.

CASE STUDY: Organizational network

When receiving a postdoctoral fellowship from certain fellowship funding bodies, recipients are often invited to attend the annual fellows' meetings to present recent work and network with other fellows. This provides an opportunity to not only present work in a peer forum, but also to develop professional connections that can persist throughout a research career.

Networking Within the Department/Institution

<u>Working groups and research centers</u>

Working groups and research centers can be formed within departments or institutions, and also between institutions, depending on where the resources are required, or where key players in a project plan are located. They generally focus on a particular research approach or method that all members employ in their research pursuits. This type of set-up provides strength in numbers when it comes to advancing the group's research, maintaining and running their core facilities, and building expertise in the field. These types of networks can often garner collective funding in a strategic framework-building capacity, and they attract staff and students to the "center of excellence."

Retreats

Retreats, in general, help to provide a concentrated focus on a topic and can promote cohesiveness in any group of people who are working together. In the academic research context, retreats can help develop professional networks within a group of people. This aids expansion of research interests, collaborative potential, and supportive professional relationships that otherwise would not have had the setting in which to develop.

Retreats can be aimed at research groups, postgraduates, postdocs, group leaders, or several groups of researchers focused on the same field or discipline. The social component of retreats creates an informal team-building dynamic that allows the groups attending the retreat to establish more developed social connections that aid professional links.

Seminar series

Seminar series can be arranged by departments, research centers, interdepartmental focus groups, or individual research groups. In any guise, having the opportunity to present your work to a collection of colleagues is an outstanding approach for networking yourself. In addition, it presents an excellent opportunity to practice your presentation skills in preparation for larger conferences and meetings. Students and postdocs who are required to speak every few months can use this opportunity to prepare new results and outline slides that they can roll over to larger presentations.

Groups and associations

Some larger institutions host associations that create their own activities to promote networking within a particular peer group in the research institution. This is typically seen at the graduate student level and/or postdoctoral level. These groups and associations can organize events, links, and resources for participants to promote their networking, or simply instigate social gatherings to encourage networking. This can create opportunities for researchers at this level to compare notes regarding their progress and help each other by sharing tips regarding approaches they can take in their work, and it provides opportunities to learn about the larger selection of resources within the institution — resources that they may not have been aware of.

Department of Biology
VOLLEY BALL
COMPETITION

2PM - 5PM

Friday 23rd May

University Sports Field 1
Refreshments Provided
BBQ to follow

Submit teams of 6 to J.Briggs
by Thursday 22nd May.
jbriggs@uni.edu

Social events

Social events within a research institution are absolutely essential for promoting cohesion and professional networks within the institution. Academics can be prone to sticking their heads down and focusing on their research, or on the literature, and never emerging. The workload is so intense. Regular social events at an institution can promote institutional connections that encourage better links between groups. These might include: wine and cheese nights, sporting events, barbeques, traditional events (specific to local cultures), the celebration of good publications, birthdays, or simply weekly gatherings in the associated staff and social club. Being professional does not exclude professional social gatherings.

Training workshops

Training workshops are not only sources of professional development for learning a new skill or honing one that you already have, but they are also **hotbeds for networking**. In fact, you will network very effectively with colleagues who are performing many of the same tasks as you. This presents an outstanding networking opportunity to develop links with people who can help you troubleshoot in the future, or who can potentially become future professional collaborators.

Coffee mornings

Some departments organize regular coffee mornings where cake, coffee, and tea is supplied. These have the capacity to draw researchers from the woodwork, usually seeking cake, but it also presents an opportunity to develop social networks within the department. These might also relate to celebrations for important events, such as: anniversaries, events, funding success, journal publication, or another special event.

An overall summary of networking pros and cons is outlined in table 5.2. This provides an opportunity to consider how different approaches might prove useful.

Table 5.2: Networking Pros and Cons

	PROS	CONS
Conferences	Meet, network, communicate, present, expand knowledge.	Travel, time away from research, cost.
Summer schools	Network, communicate, expand knowledge, skills.	Travel, time away from research, cost.
Direct inquiry	Meet, communicate, source resources.	No response to inquiry.
Referral	Meet, expand knowledge, source resources.	Reliance on third party.
Organizational networks	Meet, network, easy access, communicate, present, expand knowledge.	Inefficient communication, unreliable network.
Digital networks	Network, easy access, communicate, source resources, expand knowledge.	Inefficient communication responses, unreliable network.
Retreats	Network, communicate, source resources, expand knowledge.	Time away from research, cost.
Seminar series	Source resources, expand knowledge.	Predominantly passive.
Social events	Network and communicate.	Typically informal and potentially off topic.
Groups/ associations	Network expansion, extra resource access.	No benefit, inactive network.
Training workshops	Meet, network, communicate, present, expand knowledge.	Time away from research, cost.
Coffee mornings	Meet, network, communicate, caffeine, sugar, free cake.	Caffeine and sugar.

Promoting Incentive

You might consider offering a financial incentive, but it can be said that most academic researchers do not get into their field for the money. Thus financial incentive, whilst often effective, provides generally short-lived satisfaction to someone striving to achieve professional goals. Most academics are buoyed by producing outstanding work, which can be a relentless roller coaster of ups and downs, and twists and turns. A range of incentive-promoting approaches are discussed here.

Institutional awards

Academic institutions are renowned for their capacity to rank according to excellence, and this is no exception in the postgraduate or professional space. The provision of awards relating to exceptional productivity in research, contribution to the institution's profile, winning awards in teaching, and achievements in postgraduate study are just some accolades that are typically offered within an institutional space. These promote **the pursuit of excellence** in the academic research space, and can be combined with extra activities relating to celebrating the awardees. This creates a sense of community and pride in professional pursuits.

> **TIP: Companies from which the department purchases large resources may be petitioned to endow an annual award for researchers available to department members**

Grants, fellowships, and scholarships

Writing competitive grant applications and designing excellent projects is an essential component of successful academic research pursuits. This excellence can be developed through scholarship and fellowship programs for junior researchers. To "win" a grant, scholarship, or fellowship in your name is something that not only provides monetary reward, but also confers prestige that can carry on to promote career progression and opportunity.

Professional opportunity

A wealth of professional opportunities is an incredible motivator to being proactive in pursuing your research. For PhD students this constitutes being aware of numerous opportunities for future employment options. For postdocs this might constitute funding opportunities to extend their position, links to industry, opportunities for tenure-track positions, or access to excellent institutional networks to support their professional growth. For group leaders this might constitute access to numerous resources or cross-disciplinary networks to expand and evolve their research.

Travel

Travel for research collaboration, conferences, networking, or workshops represents part of the appeal of being an academic researcher. Similarly the freedom to live and work internationally, with exceptional mobility potential, provides tremendous incentive for some researchers.

Academic freedom

Whilst some academic positions can represent good financial rewards, the greatest appeal for pursuing this direction in research is often the freedom of professional direction that comes with working in the academic research sector. Nonetheless, with an increasingly competitive research sector bidding for dwindling funding, academic freedom may require trade-off at times in the form of industry collaboration, service provision, or consulting, for instance.

INCENTIVES PROMOTE THE PURSUIT OF EXCELLENCE

Mentorship and Support

From the group leader's point of view, mentorship should be given to all members of the group, with particular focus on postdocs and graduate research students. But mentorship can expand beyond the group setting to overall professional mentorship and support, also at the senior level. In fact, mentorship can span departments, institutions, and national and international networks. All participants in the research game can learn something from someone who has already achieved the professional goals that they are currently pursuing.

Postgraduate mentors

The **classical mentoring relationship** in academia is that of the student and their supervisor/mentor. This traditional academic partnership is becoming more challenging to achieve as academic life evolves. This is true in particular with the advent of very large research groups, in which the group leader cannot possibly mentor in a classical apprenticeship model. The pressure of producing high-quality research outcomes whilst managing teaching load and departmental responsibilities has made the role of the group leader incredibly intensive.

Extra mentorship support

New approaches to mentoring graduate students are evolving. Departments are promoting official **joint supervisor** or primary/secondary mentoring arrangements, in which students are officially supervised on their projects by two separate research leaders. Students may also be directed under sequential supervisors at different project stages. This provides extra support for the student and more opportunity for group leaders to be recognized as supervisors, despite it being only an adjunct role for one of the group leaders. Students being mentored this way can gain support from leaders pursuing diverse research foci.

111

Institutional support

Encourage your students to find an institutional mentor outside of your research group. Effective mentoring approaches can help to combat the feeling of isolation and entrapment that students and staff often suffer when they answer only to one group leader.

Extra support is now more commonly being offered by **postgraduate schools and departments**, which offer courses and workshops to help with writing and presenting skills, and which also offer research management guidance and postgraduate retreats. Mentorship is often offered by professional postgraduate school staff, and network building with peers is strongly promoted. This type of support helps the project mentors to focus on the projects at hand, because they can encourage their postgraduate students and staff to build their capacity through the institutional networks and training support provided by these extra support methods.

Nonetheless, many universities fail to take full advantage of the resources that they have at their disposal. Senior academics and group leaders are often pressured to manage all postgraduate training in the traditional approach, which encourages mentor/mentee relationships. This overlooks the very real commercialization of research and the pressure to perform on an international stage, which is nowadays commonly encouraged of academic staff at a university. By removing the pressure of teaching graduate students basic skills on a 1:1 basis, institutions can indirectly promote more productivity from their research groups.

POSTGRADUATE SUPPORT PROMOTES RESEARCH PRODUCTIVITY

Postdoctoral mentors

There is an increasing trend to support postdoctoral professional development. However, this has not been a traditional focus within the academic sector. Time spent in postdoctoral training is being extended as the sector becomes more saturated with skilled researchers and insufficient tenured positions to support them. Mentorship programs are becoming more common within institutions to help postdoctoral fellows or early-career researchers define their professional direction and goals as they progress after graduating from their PhD. Given that far more students graduate from doctoral studies internationally than there are tenure-track positions, identifying personal/professional direction for postdoctoral fellows currently working in institutions can significantly contribute to individual research productivity and drive to achieve. This is especially true when individuals know that there are good professional opportunities in sectors outside of the classic tenure-track system. Postdoctoral retreats, virtual networks, and mentoring systems are becoming more commonplace in institutions worldwide.

Tenure-track professional mentors
Once you get to the position of group leader, you find yourself confronted with extensive responsibilities in excess of anything you may have confronted as a pure research-focused professional. Juggling the requirements of a tenure-track role can be overwhelming. Some approaches that can aid professional development at this level include:

- **Mentoring programs,** in which group leaders buddy up and collaborate in supporting each other's work.

- **Research centers,** often within institutions, that help groups of researchers develop and direct research in a common focus area.

- **Training workshops,** which are focused on helping group leaders become better managers of research, people, and budgets; and on being better educators — both in the research and undergraduate coursework space.

Buddy systems

An approach that is commonly seen in businesses involves the assignment of a "buddy" to help a new member of the group feel welcome. The buddy is meant to provide assistance to the new person, helping them to feel at home and to acclimatize themselves to the way things operate in your institution. Usually the buddy will have been in the institute for some time and will be able to help the newcomer integrate smoothly into the operations of the group. The buddy should be selected carefully, ensuring that there is minimal overlap between their work and that of the newcomer. This might mean that new students or postdocs are teamed up with people from other groups within the institute via a postgraduate/postdoc network or social group, or via departmental coordinators, upon induction.

Institutional Counseling / Support

Professionals focused on supporting postgraduate academics at all levels are being employed with increasing frequency at universities internationally, in order to provide support for all aspects of academic research and work. A broad list of institutional support is noted in chapter 1, but those specifically promoting professional development include the following.

Teaching and learning professionals are on hand to help lecturers develop their course materials, and to determine the best technologies and approaches with which to deliver lectures.

Management and education staff are on hand to aid the professional development of senior researchers in their capacity of managing all aspects of their research group, including budgets, resources, staff, students, and facilities.

Health and wellbeing professionals provide counseling support, which should be backed up through institutional networks.

Careers advisors are available to aid professional development approaches, and to provide advice regarding professional direction and, more practically, specific applications. Careers advisors often organize talks and opportunities with industrial or government employers looking for candidates in your field.

Professional coordinators are often employed to provide overall support for a body of postgraduates, to aid their overall development and ensure the program runs well for all.

Safety and ethics committees should be on hand to provide guidance on safe practices — not only in experimental pursuits, but also in the day-to-day workspace.

END OF CHAPTER 5 SUMMARY
Networks and Professional Development

In this chapter we have discussed ways in which you can develop professional networks both external to and within your institution. Similarly, we have covered ways to incentivize researchers, and the provision of mentorship, counseling, and support. These key messages included:

- The entire group can benefit from different approaches to networking, and networking can be both internal and external to the institution.

- A variety of incentives can promote excellence in research pursuits.

- Mentorship can enhance professional development at any academic level, and several strategies were discussed.

- Institutional support is often available to aid professional development and professional practice at most academic institutions.

REFLECT

1: Have you been effectively networking at conferences to develop a professional network with peers and research leaders? Do you prepare well prior to attending? Do you train your team in this regard?

2: Do you seek out networks and participate in professional development through summer schools, direct inquiry, or other means? How could you improve on your professional networking?

3: Does your department or institution host actively organize networking opportunities?

4: Do you and your team take advantage of professional development opportunities? How could you improve on this?

EXERCISE 5.1: NETWORKING WITHIN INSTITUTION

Consider the following table outline, and complete it regarding the current networking opportunities that are in place at your institution. What is working? How could it be improved? What do you think your institution could benefit from? What networking have you experienced as a junior researcher that worked well?

Networking opportunity	Outcome/effectiveness	Improvement?

EXERCISE 5.2: PROMOTING INCENTIVE

Draft a list of the awards that are available to you, your group members, and your team. Include cut-off dates for applications to relevant awards, and specifications for achievement. Keep this list up-to-date as a resource to encourage your staff and students.

Award	Due dates	Requirements

EXERCISE 5.3: MENTORING AND SUPPORT

Consider the list that outlines your current group members from exercise 2.1. Use this outline to identify the types of mentoring and support that each one of your group members is currently offered. In your next individual meetings, have a discussion with each of your team members regarding whether they feel that they have sufficient mentoring support, and discuss approaches that might improve their access to mentoring opportunities.

Group member	Current support	Potential support

Chapter 5 — Downloadable Materials

Download from www.practicalacademic.com

- Exercise 5.1: "Networking within institution" — Excel file containing a blank template and an example of a completed exercise for reference.

- Exercise 5.2: "Providing incentive" — Excel file containing a blank template and an example of a completed exercise for reference.

- Exercise 5.3: "Mentoring and support" — Excel file containing a blank template and an example of a completed exercise for reference.

CHAPTER 6

Organization and Time Management

Many of the concepts covered later in Section 2 (on project management) will be applicable to overall research project planning. However, this will be touched upon here first with relation to scheduling your team, your meetings, and your individual junior researchers, because better scheduling improves overall group functioning. If your institution uses a calendar system available through the institutional computer system, encourage your team to use it to establish your time schedules.

Whole Group Scheduling

Regular group meeting schedule

A regular meeting should be scheduled each week as a set block on every group member's weekly calendar. The timing and duration depend on the dynamic of the work being pursued by the group and on the number of members who are required to attend, and the timing may also be open to adjustment depending on the dynamics of any given weekly activities. This group session could be adapted in a multipurpose fashion for the group to meet in a variety of forum styles.

Round table: Typically fortnightly, the group discusses their work in a round table forum, bringing up any significant issues relating to the group as a whole or to their particular project. This will include discussion about logistical issues and day-to-day group management, and each project's progress could be updated in brief to address any concerns that can be quickly discussed. Any general business can be brought up in this discussion.

Journal club: In order to encourage all group members to keep up to date with the most recent state of the art, and to support each group member's capacity to critically assess a research article, regular group journal clubs are essential. This also provides opportunities to discuss potential new technology or approaches that address the group's major research questions. In particular, this type of forum provides an excellent opportunity to mentor junior group members and develop their skills in literature review.

Project presentation: Including regular meetings that take the format of a formal presentation provides a great opportunity for group members to hone their skills in oral presentation, which are required for conference talks and speaking in other fora. Similarly, a rotating talk schedule facilitates regular opportunities for the group to give projects feedback and discussion, to troubleshoot project-specific issues that may have arisen, and to put forward approaches that may currently be in the planning stages.

<u>Proposed schedule</u>
Each four-week period might follow this schedule:
week 1: single presentation (general discussion prior);
week 2: round table, general business;
week 3: journal club (general discussion prior); and
week 4: round table, general business.

This way, each group member is able to present their work to the group every few months, giving them time to develop a presentation in that time. Thus when the time comes to present at a bigger meeting or thesis defense, they should have a collection of slides in their toolkit from which they can select and that are specifically relevant to their current work. Similarly, their presentation skills will be honed for communicating well in a more public forum.

Contract-wide Scheduling

The process of establishing the individual schedule for each group member or project team will be largely defined by the projects that the individual or team is pursuing. How to effectively schedule a project or research training time is probably best discussed in the context of project plans, which is in Section 2 of this book. Nonetheless, a broader schedule can incorporate the overall expected activities peripheral to the projects underway. A professional schedule outlined in a GANNT chart, indicating overall goals for the staff/student in question, will help to guide the research direction of a student or staff member at the beginning of their role. Issues that should be discussed regarding their overall schedule include:

- individual training requirements,
- annual leave,
- conferences and meetings,
- collaboration trips,
- teaching/tutoring commitments, and
- departmental activities.

A model GANNT chart is shown on the following page (table 6.1) outlining a basic two-year professional plan for a postdoctoral fellow. Two low-risk projects constitute the fellow's main output, whilst a collaborative project with a PhD student should yield a more substantial output which might extend beyond the timeframe of this fellowship. The fellow's teaching commitments, annual leave, and expected deliverables are outlined. Individual training, collaboration trips, and departmental activities are not noted, but may arise during the course of the fellowship.

Table 6.1: Contract plan - two year postdoctoral contract

	1	2	3	4	5	6	7	8	9	10	11	12	13	14	15	16	17	18	19	20	21	22	23	24
PROJECT 1 — mutation study																								
Set-up, ordering, and practical plan	▓	▓								▓														
Cloning		▓	▓																					
Transfection			▓	▓																				
Bioassay and binding assays								▓																
Protein expression and purification					▓	▓																		
Mutant protein binding assays						▓	▓																	
PROJECT 2 — support PhD																								
Help design stage	▓							▓	▓			▓	▓	▓	▓	▓	▓	▓	▓	▓	▓	▓	▓	▓
Cloning										▓	▓													
Stem-cell culture														▓	▓	▓	▓	▓						
Injection and animal husbandry								▓																
PROJECT 3 — signaling study																								
Set-up, ordering, and practical plan										▓	▓												▓	▓
Cell culture and sample collection											▓	▓	▓	▓										
Mouse work and sample collection														▓	▓	▓								
Microarray and analysis																	▓	▓						
Northern and western analysis																				▓	▓			
Annual leave												▓							▓				▓	
Lecturing/teaching			▓	▓										▓	▓									
Conference														▓										
Publication									▓												▓	▓		

Individual projects   Collaborative projects   Lecturing/teaching   Annual leave

123

Short-term Scheduling

Delivering your work on time is a critical component of success in any professional undertaking. This is usually discussed relative to delivering projects within set timeframes; however, more short-term time management of a research undertaking is essential for success.

Weekly schedule planning will come down to the individual researcher, but a typical week may incorporate several project foci, training commitments, and meetings. The contents of an individual schedule will relate predominantly to the role within the research group and the related responsibilities. Similarly, the regularity of meetings between teams and any supervisory meetings must be included in the schedule.

Supervisory meetings

It is wise to encourage researchers to provide an overview of their time schedules during supervisory meetings. If you mentor your staff and students to become effective time managers, it gives them a positive skill to add to their professional skillset.

The scheduled work that will take place over a defined period should be openly and regularly discussed between the supervisor/group leader and their students and staff. The checklist outlined below (table 6.2) provides an outline for the content that may contribute to a work schedule for the following weeks or months.

The intent here is not to ensure that the staff member is busy, because busy-ness is assumed given the extensive amount of work required in a research position. Helping your staff to ensure that they are scheduling well and not overlooking aspects of their role is important. However, keeping track of staff who may be reaching burnout and might require extra assistance to complete tasks is also critical. Creating regular weekly/monthly schedules enables the definition of realistic and reachable work goals, whilst incorporating professional development into work schedules.

Table 6.2: Checklist for scheduling meetings

Project time	Outlining specified work being contributed to each project over the next 2–4 weeks.
Meetings	Specify expected meeting schedule over the next 2-4 weeks for: - supervisor, - project/team, - research group, - department, and - user groups.
Teaching	Teaching commitments.
Training courses and workshops	Institutional and external courses and workshops that you will be attending during the coming time scheduled.
Attendance to seminars	Institutional and external seminars by local and visiting speakers are a regular event to schedule on an academic calendar.
Personal time	Scheduled time off or required personal time.

Daily Scheduling

Multitasking

Multitasking is an essential skill in the life of any research academic who performs experimental work, particularly those in the sciences. The experimentalist will regularly be faced with gaps between experiments or during incubation times that can be utilized, but how much can be done during these periods of free time is dependent on your organizational skills.

Keep a "to do" list

If your team are pursuing practical research projects, each week they will likely have a collection of experiments that need to be set up and run, data that needs to be collected, and analyses to be performed.

Nonetheless, a number of smaller tasks might accrue that might be addressed in the time between experiments. Your team should be encouraged to use these small breaks to refer to their "to do" list. This list should carry a collection of tasks that are pending, but which do not necessarily require a huge amount of time. The list might comprise simple tasks like making buffers, ordering, collecting materials from the institutional store, checking or writing a work email, printing materials, sourcing a journal article for the journal club, or booking facilities that will be required for later stages of the work. If your discipline involves long trips for field work, then taking a collection of journal articles to read whilst on trips represents good planning for the researcher to keep up to date with the literature. This can represent time well spent in catching up on your state of the art in lieu of leisure reading to while away the hours.

CASE STUDY: Time management

You've employed a new technician and asked them to complete a number of tasks during their one-year contract. You are keeping track of their progress, but over the first two months they seem to be only reporting on the first major task that is nearing completion. During your weekly meeting, you ask your technician what is happening with the other tasks that were assigned, to gauge their level of completion. The technician looks shocked and replies that they only complete one task at a time and will commence the rest of the tasks once this one is complete. You are in shock given the amount of time that you can see has been wasted over the last months. How do you address this problem? You must train this technician on how to better manage their time. This might include sending the technician on a time management training course, working together to outline the expectations of a normal daily schedule, or assigning another group member to work with them to demonstrate the best practical approach to achieving the working goals day-to-day.

Record experimental times required

Experimentalists can record the **time gaps in regular methods** in order to better multitask during their day. Incubation periods of one hour can be used to set up or manage other experiments during the day. Similarly you should note on the protocol whether an incubation period must observe a set time or if it can be accelerated or left longer or overnight. This facilitates the best possible time management.

Planning for Disruptions and Delays

There is a great likelihood that you will encounter delays in projects as you pursue academic research. This is particularly relevant for those working in scientific disciplines that involve a strong focus on natural systems. Equipment will break, facilities will suffer disruptions, experiments will become contaminated or not yield the result you expected, the wrong reagents will be ordered, or key contributors to the project will suffer unexpected delays, illnesses, or personal catastrophes. The capacity to navigate disruptions and delays is a hallmark of an experienced and well-trained academic researcher.

The best researchers are effective troubleshooters, but how can they address the inevitable delays that will be encountered in their research endeavors? Quite simply, skilled researchers are well-practiced at **anticipating where issues may arise** in their research progress. In this way, good researchers can allocate any extra time in their schedule, resources, or budget that may be required to work around problems. This might be achieved by mentioning **potential individual issues** in project plans or professional timelines; alternatively, **extra time** may be allocated to a research pursuit, to allow for any unforeseen problems.

Dealing with unexpected issues *after* they occur can be approached in several ways. As discussed in chapter 4, it is always advisable to have **soft money** available to address **extra project expenses**. Similarly, you might compensate for some issues by contributing extra manpower to a task, or by outsourcing a portion of the work to a professional service provider — for example: antibody manufacturers, DNA sequencing, transgenic mouse production, or other services related to your work.

Figure 6.1: Dealing with disruption and delay

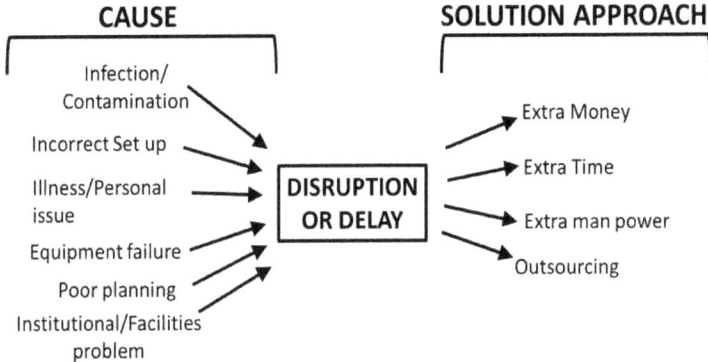

This model does not cover the kinds of disruptions and delays that may direct a project or research group into a completely new research focus, but instead outlines the approaches typically taken to keep the work on track. As you can see, many things can cause disruption or delay. Regardless of the complexity of the problem, it will require one or several of the solutions outlined above.

Regular Reporting

Encouraging your staff to provide regular time reports can ensure that they remain on-track and more aware of time passing as they pursue their work. The types of reports that you should encourage regarding projects will be outlined in Section 2. However, overall candidature or fellowship reports are separate and encouraged, to ensure that you keep up-to-date with your students and staff as they pursue multiple projects in their workload.

Regular reports might include schedules as outlined earlier in this chapter (i.e. supervisory meetings), in addition to recent deliverables or up-to-date summaries of the recent progress they have made with each individual project they are pursuing. The regularity of these reports is up to the supervisor and staff/student, but **meetings every one to two months** with reports is advisable. This not only makes it easier to track the work, but it also trains the students and staff for potential employment in government or industry at a later stage, promoting a culture of accountability and a practice of reflecting on advancements made in a given time period.

Some institutions also promote reciprocal feedback during annual performance reviews from the staff and students to their immediate supervisors. This approach, receiving feedback from those you are managing, represents a conscientious and proactive approach to improving on your own leadership skills, and may encourage your own professional development and performance improvement moving forward.

Performance Reviews

Depending on the type of positions you offer in your group, you may be required to manage performance reviews. These usually take place annually, and may be used to gauge promotion prospects or salary review. As a tenure-track academic research leader, performance reviews will also be required to gauge *your* overall work achievements during a given period of time — typically annually. The **provision of regular reports** by your team, and ensuring that you keep track of your own activities as a group leader, is key to being able to report effectively on the performance of your group over time, and on your performance as a research leader.

Balancing Daily Activities

How you balance your work and life to operate effectively will depend on your current personal situation and any external commitments in your day. A student might be predominantly focused on completing their studies, whilst a more experienced researcher might have a marriage and/or children that represent their principal personal focus outside of work. Balancing your schedule to fit your personal life will positively impact your productivity and personal happiness.

129

Accommodating Diverse Groups

If your group works in a variety of research topics, and if field work or working off-site is common, you should still arrange routine short meetings to touch base via Skype or Zoom, with more organized meetings at least monthly or bimonthly. The isolation that comes from working remotely is only compounded when you do not feel that you are a part of your larger research group. Sadly, this is a very common problem encountered by postgraduate students and researchers who work offsite. Similarly, many of your students may be challenged by very short deadlines and intensive workloads. These pressures can cause immense stress. Thus it is critical to accommodate the working style that promotes the best productivity with minimum issue for each group member.

In research pursuits that require the use of expensive and highly-utilized core facilities, many junior researchers prefer to work at night or early in the morning in order to push the advancement of their work when the equipment is available. As such, the group leader should maintain lines of communication during periods of high productivity, but should demonstrate flexible methods that might embrace more digital approaches, like messaging, email, or the provision of short reports via a shared drive during the intensive research periods. However, it is critical to ensure that you remain in communication with your junior researcher, rather than "let them go" whilst they are working through intensive practical research periods. **Demonstrating interest, providing feedback and direction, and promoting productivity** are key tasks for the group leader at all stages during a research investigation.

Work effectively as a team

The emphasis in academic research should be on ensuring that the project continues to move forward. A classic within-group team approach would include instances where some project participants work late to have more extensive access to facilities without disruption or competition for bookings, whilst more established researchers might maintain a regular work schedule during business hours. This approach is particularly effective when time-dependent work can be managed by several researchers in a team approach (chapter 2), whereby the work can be pushed even further with collaboration between team members. Tasks can be assigned during team meetings, and timings should overlap each day by at least several hours so that team members can coordinate their work, and can discuss issues and findings that have come up during the course of their work.

CASE STUDY: Team approach

A team working on protein binding and bioassay studies was conducting a combination of overnight bioassay experiments and binding assays to test the efficacy of a battery of test proteins that they had created. The team comprised of a postdoctoral fellow, a PhD student, and a technician who worked together to conduct the studies. Working all together in the mornings that assays were set up allowed them to test more test proteins at a time, and it resulted in excessive amounts of data by mid-week, providing data for analysis on Thursday and Friday. Tiered scheduling of the postdoc and student allowed them to collect time points in the evenings and occasionally on weekends to maximize the time points collected from the assays they had established each week. After set-up, the technician was tasked with ensuring that the facility remained well stocked, infection free, and maintained. Together this team produced an extensive amount of data, yielded numerous publications, ensured continued funding of the group, and boosted all three careers.

END OF CHAPTER 6 SUMMARY
Organization and Time Management

In this chapter we have discussed ways in which you can schedule at several levels, including regular group activities, contract-focused activities, regular meetings, and day-to-day activities. We also address how to confront change and manage unexpected disruptions. The key messages of this chapter include:

- Establish a stable meeting schedule that supports the needs of the group with regard to advancement, training, and development.

- Be proactive in inspiring your group members to establish clear schedules during their appointments to your team, both over the long-term and short-term.

- Try to anticipate issues that may arise in the work being pursued by your group, and encourage your people to do the same.

- Encourage solid troubleshooting skills in your group members.

- Ensure that regular reports are provided on the current work and its advancements.

- Promote effective and balanced work/life practice amongst your team.

- Try to accommodate diversity and flexibility in work practices for your team, with an emphasis on efficiency and adaptability.

REFLECT

1: Do you follow an established scheduling system with your group members?

2: When regularly meeting with your group, do you promote their communication and analytical skills through a program of regular journal clubs and the presentation of their work to the team?

3: Do you accommodate your students and staff regarding their personal situations or working requirements in order to maximize their productivity?

EXERCISE 6.1: MEETING SCHEDULES

Draft a meeting schedule outline based on the table below. Identify the regular monthly meetings that you anticipate with your group, and what structure these meetings will take. Identify whether a schedule is expected for the different meeting types each month: order of presentation, order of hosting, and order of delivering a talk.

Day/time/regularity	Type of meeting	Participants (initials)	Schedule notes

EXERCISE 6.2: REPORTING

Look at the table of your group members that you completed in Exercise 2.1 and consider how frequently you request reports from each of them. Make a formal summary of how often you require reports from these group members going forward, and in what form they should be presented. This may vary from person to person, depending on the type and duration of project.

Group member	Report types	Report details

EXERCISE 6.3: PLANNING FOR DISRUPTIONS OR DELAYS

Create a summary of the potential disruptions or delays that might impact the projects currently underway in your group, and forecast approaches that might address these potential problems. Include backup plans should key equipment fail, facilities be shut down due to contamination, or similar.

Project/person	Possible disruption/delay	Approach to manage

Chapter 6 — Downloadable Materials
Download from www.practicalacademic.com

- Exercise 6.1: "Meeting schedule outline" — Excel file containing blank a template and an example of a completed exercise for reference.
- Exercise 6.2: "Reporting" — Excel file containing a blank template and an example of a completed exercise for reference.
- Exercise 6.3: "Planning for disruption/delay" — Excel file containing a blank template and an example of a completed exercise for reference.

CHAPTER 7

Personal and Professional Dynamics

Academia is a competitive industry, and as such there are a lot of people who would sell their grandmother to get ahead. When you consider it, there are a range of reasons for this: the stakes are high; it is a good career; you can achieve recognition and accomplishments to be proud of; you can win awards; the salary may be good; and there is freedom from the normal old nine to five. Not to mention the fact that at a high level, you are essentially paid for thinking up and proving new ideas. It is a competitive creative industry that works converse to the drone-work focus of the modern era. This chapter is focused on discussing the social politics of academic research and how you might navigate it effectively whilst maintaining personal welfare and health.

Promoting Work / Life Balance

Junior researchers often work all hours to achieve their research goals, focusing entirely on career success. This is a surefire recipe for burnout, but group leaders usually encourage it, rejoicing in the well-oiled machine that is their staff working toward more scientific advancement in their name. This model has no longevity, and many have called for a reorganization of the academic career structure to better reflect real life. If research leaders are savvy in their group management, they can forego the requirement for their junior team members to burn out before they reach tenure. If you are managing a group, the best thing you can do is **establish a clear goal structure** with your staff and students to ensure that they manage their time and their project goals effectively. Throughout this book, we discuss various approaches that you can take as a group leader to promote your team's productivity and efficiency in their research pursuits.

<u>Some approaches to promote work / life balance as a group leader</u>

Lead by example — demonstrate your personal and professional organization with respect to work/life balance.

Promote progress — consider the best way to maximize your group's productivity through effective management, networking, team organization, and guidance.

Leave nobody behind — ensure that you direct projects effectively, so that nobody is left floating and unsure about their research direction and expectations.

Keep track of projects underway — make sure that you keep tabs on your people. If you are a research leader, it is like parenting; you have to give them some freedom, but gentle guidance is required.

Remember you manage humans — ensure that your people allow themselves some personal time. This is only possible if projects are managed well, and that will only occur if you are a good mentor and if you involve yourself in the management process.

Be available — make yourself accessible to your staff and students. An open door policy promotes a more cohesive and affable research environment in which staff are not afraid to approach you regarding any issues.

CASE STUDY: Work/life balance

A mid-career research leader demonstrated a very controlling hand when managing his research staff. His office was situated directly facing the the laboratory. He kept a close watch on each staff member's time of arrival and time of departure. Although he was an established researcher, he continued to work long hours — he regularly worked into the evening. Whilst staff and students were in the laboratory, he would listen in to their conversations, coming out to join in when they were of interest. The staff always felt as if he was watching over their shoulders, and were relentlessly unsettled in their work environment. The group leader failed to create a supportive environment, but rather created an industry-like clock-punching atmosphere where the advancement of research was impeded by his focus on appearance, rather than on progress.

Alliances

Different alliances tend to develop in academic research groups. Such alliances may be related, for example, to common professional goals, shared anger or disappointment with challenges coworkers are facing in the workplace, or simple social connection. These alliances can have a dramatic impact on group dynamics and productivity, which may represent a positive or negative influence. Strong connections made as colleagues in one group can roll over into long-term collaborative efforts over an entire career, given the formative experience of working in academic research.

Conversely, you might have a completely cohesive, happy, and productive research group already working together, and the addition of one person who "stirs the pot," so to speak, can disturb the veritable utopia of academic pursuit that you have established.

Schoolyard approaches to work are a hallmark of unprofessional colleagues, and as such the best approach to dealing with this issue is the recurring theme we've seen already: remain professional, follow solid guidelines, and keep focused on what your own professional goals are and do not allow yourself to get distracted from them. If you are operating well under the guidance of a superior or mentor, continue. As a group leader dealing with an obvious disturbance in your group, if you follow the guidelines outlined in this book, you are definitely on the way to maintaining your research group without the issue of unsettling social situations that can arise from unholy alliances.

REMAIN PROFESSIONAL AND KEEP FOCUSED ON YOUR OWN GOALS

Preferential Treatment

Many group leaders favor particular people or particular projects in their research pursuits or work environment. The fact of the matter is that you will have stronger, more positive experiences with some members of your team. There will be some people that you will potentially gel with more easily. It is critical, however, that all members of the team feel valued, and that each receives the same amount of management and guidance in the pursuit of their professional goals under your mentorship. You have a responsibility for your team and should support their professional development and progress on an individual basis.

Appreciate the People You Have

Many of the dynamics described above can be avoided if a project leader or group leader takes the time to focus their efforts as a mentor. Evaluating staff and students, whilst making the effort to appreciate their individual qualities, can significantly help team members to improve in all aspects of their work.

When you welcome a new member into your group, you should sit down for a discussion about who they are and where they are aiming to go from this position. There is a checklist of some questions that you could ask new staff to get to know them better in table 7.1 on the following page. This might constitute an informal discussion if it is taking place after they have been hired. It is worthwhile revisiting this type of questionnaire each year to gauge how your team members' perspectives have changed when you meet with them for an annual review. This way you can tailor your support to align with the goals of the staff you are supporting. The aforementioned list of questions can be used to frame the discussion.

Conflict

Research groups are typically comprised of an incredibly diverse collection of people with varied experiences, ideas, backgrounds, and approaches. Further, researchers typically operate under high pressure to publish or perish. Taken together, the likelihood of conflict is high. Finding the best approach to facilitate conflict resolution will depend on your personal approach. Your department may have guidelines regarding effective conflict resolution, and to some extent the appropriate conflict resolution is relative to the type of conflict and how it escalated. Finding the core problem through communication and discussion, whilst encouraging each other to listen with an open mind, is key.

In dealing with conflict, group leaders must take an impartial and non-condescending approach to negotiating conflict resolution, remembering not to belittle any parties regarding their point of view. If one of your group members has a problem with another colleague, it is worth your while to calmly determine the root cause of the problem in order to first solve it, and to then ensure this does not happen again. Sometimes it is the culture or atmosphere established in the lab through the management of the group leader that encourages conflict in the first place.

Table 7.1: Personal goals and characteristics questionnaire

Name
Position name and description:
What are you hoping to accomplish from this position that you are commencing?
What do you think the challenges in the role will be?
What do you think a satisfying aspect of the role will be?
Do you like collaborating/working with others?
What challenges have you faced in your previous professional experiences?
What do you personally dislike in the workplace?
What aspects of this job do you like the most/least?
What is important to you in life? Family? Balance? Professional direction?
How do you like to work? Hours? Mornings or nights?
How would you like to develop professionally?
Where do you see yourself in five years?
Would you like to attend meetings/congresses?
Would you like to participate in teaching/mentoring?

Group members that are embroiled in conflict with their colleagues spend far less time focusing on productivity, and far more time focusing on negative personal issues that could be avoided if the group was managed better. As a group leader, you are better served promoting a cohesive workplace than promoting competition and rivalry within your group. This type of atmosphere is not conducive to productivity and often escalates to more severe issues.

Ways to manage conflict issues between research staff

There are a range of conflict types that might take place between your team members. The manner in which you, as the group leader, manage the conflict will depend on the level of conflict that you are dealing with.

- Evaluate the level of conflict and overall situation to determine how communication must be brokered.
- Gauge the reason for the conflict.
- Listen to grievances from all sides and stay on topic, without distraction and objection.
- Transform confrontation into conversation wherever possible.

LOW-LEVEL CONFLICT

Low-level conflict situations might relate to taking other group members' reagents and not replacing them, and other similar petty annoyances with each other. In these cases, the matter might be resolved fairly simply with a clear discussion and agreement of how to manage the concerns of both group members.

MEDIUM-LEVEL CONFLICT

Medium-level conflict situations might relate to ongoing conflict in which group members consistently fight or have a recurring issue that needs to be addressed. These might evolve from persistent low-level conflict or derive from an acute conflict situation in the workspace.

HIGH-LEVEL CONFLICT

High-level conflict situations arise from time to time, and can be due to a number of issues. These might include intense personal issues, racial tension, sexual harassment, or smaller conflicts that have been left unchecked and thus have escalated.

Classic approaches to addressing conflict

1: Call a meeting to discuss the matter with both (or all involved, if more than two people are in conflict) group members.

2: Keep matters clear regarding the issues being confronted.

3: Implement potential short-term and long-term solutions that should reduce the likelihood of the problem recurring.

4: Monitor the situation.

5: Bring in institutional support should the conflict not resolve.

Common Academic Conflicts / Problems

Authorship disputes

Decisions about authorship are frequently central to conflicts in research groups. Who deserves first authorship, and whose name should be included on the manuscript at all? These issues are discussed more extensively in chapter 15.

Pay dispute / envy

Researchers are often sponsored by fellowships and scholarships, and those who are paid more or receive better fellowships are often the envy of other equally hard workers who sadly missed out on such illustrious funding opportunities. Unfortunately, this disparity is commonplace in academia, and it is a part of the sector that all participants need to accept before moving forward. See the "group collaboration" case study in chapter 3.

Work disparity
Almost every poorly managed team has someone who fails to pull their weight on the project for some reason or another, which can build resentment in other team members.

Lack of clearly defined roles
When team/group roles are not made clear through regular meetings and communication, competition, instead of cooperation, takes place. This is often related to scorn or anger over some group members contributing more to keep the lab operational, whist others fail to contribute to general duties. See chapter 4's case study on rostering duties.

Arguments over resources
Equipment, appointments, limited resources, taking your colleagues' resources, and access to shared resources and tools can trigger heated arguments that persist until the issue is resolved or somebody moves on.

Obvious lack of respect for your colleagues
The supervisor should demonstrate social and professional behavioral expectations by example. However, pushy, obnoxious, and competitive colleagues can sometimes dominate the research environment.

Personal relationships
Given the amount of time that researchers spend working together, romantic relationships invariably evolve. Similarly, due to the high mobility of academic research, couples may relocate internationally together and work in the same discipline to facilitate their relationship. Relationships do suffer and change, however, and personal dramas can overflow into the workspace.

General Approaches to Reduce Conflict

Develop a clear outline of expected behavior in your research group. In this outline, state the expectations for all group members with regard to:

- Courtesy.
- Cooperation.
- Meetings and journal clubs.

- Collaboration.
- Note keeping.
- Rules for sharing resources.

You may also wish to:

- Develop a clear policy for whether you will allow postdocs to take their projects with them.

- Design projects in which your staff and students are not competing.

Table 7.2: Code of conduct

This is a typical outline of a laboratory research group code of conduct, which might be recorded in the group reference folder.

OFFICIAL GUIDELINES

Group members will read and follow the university guidelines set out as follows:

- Code of Conduct for Responsible Conduct of Research, provided by the research office of the university.
- The National Code of Conduct for Responsible Research, accessible via the policy document space online.
- The Ethics of Animal Experimentation guidelines outlined by the National Code of Practice for the Proper Care and Experimental Application of Animals for Scientific Purposes.

RESPECT

Group members will respect the individual work spaces provided to their colleagues for both office and laboratory-focused work. We will acknowledge each other in the workplace and, if any conflict arises, will meet calmly to negotiate the issue to reach a mutually agreeable solution. If the conflict is unresolvable, we will meet together with the group leader to find a resolution via mediation.

STOCKS

Group members will replenish communal stocks and resources when noted to do so:

For reagents: sterile PBS, MilliQ water, communal stock solutions — when only 2 liters remains of any given solution.

For chemicals: when the common stock solution is below 1/3 of the final bottle.

For enzymes: please see the enzyme list for more specific details regarding replenishing stocks.

For office supplies: Please replenish from the departmental store as you reach approximately the last 10% of any resource in the office supplies cabinet.

COMMUNAL EQUIPMENT

Booking equipment will take place via the sign-on sheets that are managed by the group technician, located proximal to each piece of equipment. If you wish to use the piece of equipment when it is booked, you will need to negotiate with the person that has booked the equipment for that time period. If you wish to make blanket bookings over many days, please advise the other group members at the nearest available group meeting prior to that booking period, or by email to the group, allowing for negotiation where applicable.

Cleaning and general maintenance of equipment is the responsibility of the users. Please ensure that equipment is left in the same condition in which you found it. If any issues arise with the equipment that you are using, please report them to the person responsible for the piece of equipment in question, which should be noted close to the piece of equipment itself.

SAFETY

All group members are required to follow the institutional guidelines for safety. For all standard protocols, which are recorded in the group reference materials, safety has been addressed and best approaches noted. Similarly, all group members are required to consider and plan for the most effective approach to safety when developing new experiments that are not already listed in the standard operating procedures.

MEETINGS

The group will meet at least once a month to discuss general issues in the day-to-day operations of the group. All group members are expected to attend.

RECOGNITION

It is expected that all contributors to a research project will be recognized when the work is published, patented, or submitted for other presentations. Any issues regarding contributions will be discussed with the group leader, or taken further to institutional mediators when required.

RECORD KEEPING

All group members are expected to keep clear and appropriate records of all of their experimental work. These records should be able to support patent application, should the work become patentable during the course of any investigation. No work is to be kept secret from other group members, because we operate our group in an open and collaborative style.

Ownership of the Work

Many junior researchers work in a group headed by a research leader who is typically based in an academic institution on a tenure track. These researchers often face issues when it comes to the end of their contract, fellowship, or postgraduate studies, and the future of their research comes into question. Many postdocs expect that they can move on, taking their research focus with them, so as to build a solid program that might see them lead their own group in the near future. PhD students, in contrast, may continue on in their field, or may decide to make a change in their research focus.

Problems arise when there is disagreement regarding who effectively "owns" the research once the partnership between the group and the researcher comes to an end. Group leaders want to retain in their group the research that underpins the strength of their program, and to hold the rights to take the work forward. This is usually the case when they have funded the work and conceived the work at the outset. Nonetheless, particularly invested researchers may play a very active role in directing the advancement of the work, and may have — on their own merit — earned a majority of the finances that supported the project.

For all of this confusion, it is critical that this issue is addressed up front, at the beginning of a research contract or project. The issue of who "owns the rights" to continue with a research project when someone leaves the group should be openly agreed upon. What is clear is that if a group leader repeatedly demonstrates disrespect and lack of consideration to their staff's professional development needs, they will fail to see great success as a leader in the long run. **Professionalism, clarity, and communication are key** to ensuring that everyone participating in a project is clear on the expectations of whether they can extend their work further in future positions elsewhere, and whether they will have access to the resources that they might generate during their time in the current group.

Recognition for Your Work

A common ethical issue relates to recognition of work that has been completed by a researcher who then moves on to another group with a different research focus. Although you've moved away from the group, you've left two-papers'-worth of data ready to publish, perhaps awaiting one extra figure or experiment, perhaps awaiting a collaborator's contribution. Maybe you've published your thesis, which included an incredible review of the state of the art. Perhaps you have written a grant application for your supervisor and left it in their hands, and it has been submitted or is under review. The best practice is for the group leader to acknowledge the contribution of everyone who worked on the project, regardless of whether they have moved on. Sadly, personal issues often negate professionalism in recognizing contributors, particularly once they have left the group.

EVERY CONTRIBUTOR TO A PROJECT SHOULD BE ACKNOWLEDGED IN THE PUBLICATION

CASE STUDY: Ownership of the work

After completing a five-year PhD study and seven years in the same group, a PhD student secured a well-renowned research postdoctoral fellowship in another country and left shortly after completing their PhD. The PhD student wanted to publish the extensive introduction to their thesis as a literature review, but the PhD supervisor said that they did not think it was a good idea. Six months later, a paraphrased review on the same topic was released by the group leader together with the postdoc that had taken over the project. The review failed to acknowledge or refer to the then-PhD student in any way. The student, wishing to retain their PhD supervisor as a referee, said nothing.

Ethical Misconduct

Given the competitive nature of academic research and the high pressure to routinely produce innovation, unethical conduct can take place. This might be from the group leader, staff, students, collaborators, or institutional managers. The following case study outlines one example of clear unethical conduct, but often this may be a gray area.

CASE STUDY: Project competition/sabotage

A group that worked at a prestigious university had been encountering issues in their research for more than six months. None of their experiments had been working, despite numerous troubleshooting approaches taken, and it was across the board: all group members were struggling to make any experimental headway. This was particularly an issue for the PhD student, who was trying to complete their thesis to submit later that year. The student would have to extend their candidature to finalize their project.

Soon after, the group began the move to a new facility, and for a time the group leader ran two groups between the two facilities. The PhD student, in desperation, moved to the new facility. Here, start-up money for the new group was able to support the student's stipend, despite their continued enrolment at the old facility. Within two weeks, the student's experiments started working, despite the fact that they were the first one on the ground and that they had set up the lab almost completely by themself. When the remainder of the lab team joined the new facility two months later, none of their experiments worked again.

The group leader, believing that something was amiss, installed video cameras in the laboratory to see if there was some type of sabotage. Within days, one of the group members was recorded on film, sabotaging everyone's work after hours.

Unfortunately, this type of case is not uncommon. Nowadays it is termed "Academic Sabotage" and comes about predominantly in response to highly competitive research environments. This type of incident can take place, however, even under the best of circumstances. The case described above occurred in a cohesive and friendly group who were always going out together socially, in addition to working well together, and who had the assistance of a very proactive research technician and a very supportive group leader. Nonetheless, it is important that group leaders clarify at the outset that unethical behavior will not be tolerated.

This type of misconduct can occur at all professional levels in academic research, as we often see in popular media. It can bring significant disrepute to academic institutions and the downfall of a hardfought career. Famous examples seem so disconnected from what you may experience day-to-day, but in a competitive research industry — where many postdoctoral fellows are vying for a progressively limited number of tenure-track opportunities and consistently reduced funding opportunities, and with increasing graduate numbers — it is increasingly common to see people performing this type of misconduct to achieve success or career progression.

Another classic case of misconduct due to competition is shown in the next case study, involving the **undercutting of a colleague's work** to promote one's own work. This might also appear in the guise of not replenishing resources, not sharing equipment, or manipulation of resources to limit the access of others. This can also occur at higher levels, between groups or, more commonly, among anonymous reviewers who reject a submitted paper due to their own similar work that is nearly ready for publication.

CASE STUDY: Undercutting colleagues' or students' work

A Masters student was working on the final experiments of their research project and had a limited amount of a home-made reagent remaining to complete a triplicate dataset for his thesis. A research assistant was requested to help the Masters student in the final stages of their award project. Both were expecting to commence a PhD in the new year and would be competing for funding and space in the same group. The research assistant roughly completed the experiment with very little accuracy and the resulting graph had a skew of data points that made no line of best fit. The Masters student was left with unworkable data for their thesis and no more reagent left for a last attempt to complete the dataset before submission.

How should the group leader handle this situation? Is it clear to the group leader that this is taking place? Would you notice?

General Personal Issues

Many types of personal issues can arise in the academic work space. These may include having to work with someone whom you dislike, who demonstrates terrible personal hygiene, or who treats you with disrespect. Similarly, alliances form and reform, people talk behind others' backs, and general social discontent can result from poor management. As a group leader, it is essential to mediate any disputes or issues arising from these situations, to facilitate the best possible connectivity and collaborative potential of your research teams.

Sexism and Racism

> **CASE STUDY: Recruitment sexism**
>
> A senior researcher sourced a research grant together with a group leader. The grant came with funding for a research fellow who would be working on the project. After putting out an advertisement for a postdoc, a number of replies were received. Although the research topic was quite specific, most of the applicants had clearly used generic resumes. One applicant stood out significantly with a CV tailored to the position.
>
> During the interviews, it was clear that all but one of the candidates had come unprepared; but one woman came to the interview with experience in the key focal area and two recent research articles on the topic to query, including one that the researchers had not seen before. After the interviews the group leader suggested that the only male candidate be hired, which was not the prepared candidate.

If you look back over recent media, no matter when you read this chapter, you will likely find big dramatic stories about sexism or racism in science or in academia, or both; yes, both are present, just as they are in many professional spaces. Thankfully, there are advances being made in reducing these issues in the professional academic space, but these issues continue to be a persistent concern.

Sexism happens in both directions. More senior academics may select more attractive candidates, or those of a particular sex, or may favor them throughout their contract/candidature. Similarly, staff and students may employ their attractiveness to manipulate the more senior academic staff to promote their own professional advantages. The manner in which you, professionally-speaking, deal with this issue will depend heavily on the resources that you have at your university. From the point of view of your professional practice, being aware of the fact that **you do not condone sexism in your team** or amongst your professional colleagues is key to ensuring an ethical and long career.

Although academia is a particularly international profession, professional racism also continues to occur. Group leaders may recruit people of a certain ethnic or cultural persuasion because they consider them to be harder workers, or easier to control. Similarly, multinationals may be recruited because they bring their own money, and then find that they have minimal support on the ground once they arrive at an institution. In fact, some expatriate researchers encounter such terrible racism in the environment in which they are working, or the city in which they are living, that they crave constantly to return to their home country, and find it difficult to function in their professional and personal life.

SUPPORTING THE ACCEPTANCE AND INCLUSION OF YOUR MULTINATIONAL STAFF WILL PROMOTE WORK HAPPINESS

Sexual Harassment

Sexual harassment is a common problem the world over for students. It is encountered at all levels, and oftentimes is difficult to prove. Even when complaints are made, established professional academic networks often overlook issues, and the complainant ends up being the one who must be moved to a new situation.

CASE STUDY: Sexual harassment

A female PhD student worked long hours and was dedicated to achieving her PhD. Her department always had a large contingent of postgraduate students working late into the evening, maximizing their time to get their research done.

A senior postdoc from a neighboring lab, who collaborated with her supervisor, began visiting the lab often after hours to borrow reagents, and would always check in to see how she was doing. He did this in a very collegiate way. After some weeks, he began sending her gifts and jewelry through the internal mail system, which she found uncomfortable, and she told the postdoc as much. The gifts continued, and they escalated to emails professing his love for her.

When the student brought the matter to the attention of her group leader, she was unheeded and told that she must be mad. She then disclosed the matter to her colleagues, and this information reached the postdoc who began his own campaign of gaslighting his colleagues, telling them that she was delusional for thinking he was harassing her. The PhD student was given no support and suffered whispers behind her back for the remainder of her PhD.

This case study describes a poor response by a group leader to a matter of sexual harassment brought to his direct attention by one of his postgraduate students under his mentorship. The matter should have been escalated to university welfare services, rather than leaving the student at the mercy of the gossip grapevine in her department. It is hard to believe that this type of sexual harassment still takes place in the professional academic environment. When confronting these issues, group members should be advised to:

1) Make the issue known, and talk to your immediate supervisor.

2) If nothing is dealt with, talk to your department head. If necessary, go higher.

3) If your institution has a department set aside for student welfare, talk to them.

This behavior should not go unchecked, and the perpetrators should not be supported — as happened in this case. The lack of acknowledgement of a single case of sexual harassment can lead to further issues down the track, in which the senior academic sexual predator can continue to take advantage of junior staff. This constitutes unethical behavior and poor institutional management.

END OF CHAPTER 7 SUMMARY
Personal and Professional Dynamics

In this chapter we have covered common issues confronted in managing personal and professional dynamics in the academic research sector. The key messages from this chapter include:

- Encourage your team to establish a work/life balance that can be achieved over the long term.

- Promote professionalism in the workplace and proactively confront any schoolyard behavior from your group members.

- Show support and appreciation for your team, and do so without favoritism for one researcher over another.

- Proactively address conflict situations in your group, always seeking best practice for the resolution of any issues that arise.

- Consider creating a code of conduct to clarify the accepted personal and professional behavior of the team.

- Acknowledge the contributions that your team have made to the work.

- Keep track of what is happening at the research level in your group, and be aware of and deal with any potentially unethical behavior that may be taking place at any point in time.

- Promote a supportive environment in your group, in which your team is protected and supported when sexism, racism, or harassment might arise.

REFLECT

1: How do you resolve conflict in your team?

2: Have you observed any of these issues in your academic working life?

3: Are you dealing with these types of issues right now in your team?

4: Are there other issues you are confronting in your research group that are not listed here?

EXERCISE 7.1: CODE OF CONDUCT

Develop a Code of Conduct to include to your laboratory's resource collection. Include the major points outlined in table 7.1. Table 7.1 can be downloaded from the Practical Academic Website to use as a template.

EXERCISE 7.2: OWNERSHIP OF THE WORK

Make a summary of all the major lines of investigation in your group and consider what aspects of the work are critical for you to retain in your group moving forward. Would you consider allowing a postdoc to take a line of investigation with them to start a group? Under what circumstances? Reflect on your research team outlined in exercise 2.1 and anticipate their individual professional options/projected direction. Also consider their personal and professional goals (exercise 3.2).

Line of Investigation	Project Participants	Anticipated direction after project completion

EXERCISE 7.3: CONFLICT MANAGEMENT HISTORY

Reflect on any significant conflicts that have arisen in your research career. How were they solved? How could they have been better addressed? Are there any trends in conflicts that occur over time?

Conflict	Approach	Outcome

Chapter 7 — Downloadable Materials
Download from www.practicalacademic.com

- Table 7.1: "Personal goals and characteristics questionnaire" — Word file containing a blank template.
- Table 7.2 and exercise 7.1: "Code of conduct" — Word file containing the example from the text that may be modified.
- Exercise 7.2: "Ownership of the work" — Excel file containing a blank template and an example of a completed exercise for reference.
- Exercise 7.3: "Conflict management history" — Excel file containing a blank template and an example of a completed exercise for reference.

REFERENCES – Chapter 7

Two references that provide some great insights to managing personal issues in a research lab include the following:

Cohen CM and Cohen SL. 2012. Lab dynamics: management and leadership skills for scientists. 2nd Edition. Cold Spring Harbour Laboratory Press, Cold Spring Harbour, New York.

Barker, K. 2010. At the Helm. Leading your laboratory. 2nd Edition. Cold Spring Harbour Laboratory Press, Cold Spring Harbour, New York.

CHAPTER 8

Teaching and Departmental Responsibilities

Unless you are employed in a research-only institution that frowns upon teaching duties, you will likely be responsible for delivering courses in your field of interest during your tenure at an academic institution. Most group leaders take on the role with the understanding that they have to teach, manage a research group and its staff, participate in institutional management, likely run a facility related to their work, and network internationally at meetings and conferences; all whilst developing and completing novel research projects that are competitive on a global scale (figure 1.1). Here we discuss several approaches you might consider to improve on, and maximize your benefit from, your teaching commitments. Then we discuss the issues around various common departmental responsibilities.

Sharing the Teaching Load

Your postdoctoral fellows would be well served to practice their teaching and management skills in delivering lectures on your courses from time to time. Most jump at this opportunity if given the chance, and it is within the mentoring guidelines to offer such professional development to them. Imagine that they are so successful that you could end up having a junior lecturer developing their own group within your department or elsewhere, and that person becoming a solid future collaborator. It is the stuff of empire building. For many of your research fellows, getting lecturing experience can make the difference between securing a tenure-track role in the future or in becoming a perennial postdoc.

CASE STUDY: Encouraging postdoc teaching

A postdoc indicated an interest in doing some teaching as a part of the second-year undergraduate course being delivered by their group leader. The group leader agreed to let them develop and deliver several lectures throughout the semester. The postdoc developed excellent teaching materials that the group leader was able to include in their resources. The postdoc achieved such consistent feedback from the students that they were offered a teaching position by the head of department, whereby they were given a full-time role that included teaching and research. The postdoc was then able to continue in their original group over the longer term, eventually becoming a researcher in their own right once they had secured grant money. The original group became a center of excellence in the field, and the productivity of the two people attracted excellent junior researchers and more collaborative opportunities, given their combined strength.

This case study demonstrates how effective the inclusion of a postdoc to your teaching duties can be in overall positive professional development, for both the postdoc and the group leader mentoring them.

Maximizing the Output

If you are running practical courses, **employ your postgraduate students to deliver practicals** that benefit your research. This might include: the preparation of materials, such as crude Taq polymerase that can be used in the lab, prepped in undergraduate lab practicals; analysis of your own existing data, backed up by a lab report including a literature review; or ecological collection studies, where you might expand on collection sets that you are establishing. In this way, you may be surprised to find that you can improve upon your research outputs whilst contributing to your teaching load, and you may potentially identify students with a particular interest and skillset that matches the type of postgraduate you would like to later welcome into your group.

Juggling Your Research Work with Teaching Commitments

Graduate students and postdoctoral fellows often take on teaching roles in their postgraduate life, and these skills can help them to attain tenure-track positions in the future. Once in a tenure track role, many group leaders get frustrated with the time that teaching takes away from their research pursuits. More than ever, effective time management is essential to succeeding in both areas. Some potential approaches, in particular for group leaders, are those outlined below.

Try integrating your research questions into your teaching. This can not only keep the students in touch with up-to-date cutting-edge research, but it can genuinely provide new insights into your data. Few researchers would not see the potential worth in this. Student-centered and activity-based teaching promotes deeper learning, so why not see if your students can solve some of the pressing issues that you've been confronting in your research? Practical classes also present an awesome opportunity for you to expand on your data points. There is no reason why you cannot benefit from the resources and time spent training students in the laboratory, and there is no downside to getting them to do practical things that can actually yield research outcomes or publishable material for you and your group.

Employ your postgraduate students as tutors. Most university research departments employ their PhD students as casual tutors for practical classes, providing them a part-time income and tutoring/teaching experience that they can include in their CV. If you follow the above advice and develop practical classes that complement your research pursuits, having your PhD students teaching the classes can ensure that you have your finger on the pulse; further, you will have experienced tutors ensuring that the work is performed correctly, as it would be in your group, who themselves are benefiting from a well-run practical course.

Teaching as a Researcher

Engage in pedagogical practice classes. You've likely secured your role as a group leader by being an incredible and productive researcher who can secure research funding. You've probably got some amazing qualities, which stem from all of your research experience. You may or may not be a natural teacher. In any case, if a major component of your professional role is to educate, you need to acquire skills in educating. Remember that now you are a group leader, you have a responsibility for both your undergraduate and postgraduate students. The way that you were taught is not necessarily the best. Coupled with good proactive approaches to managing your research, good pedagogical practice can set you apart from the rest, and it can bring opportunities you had not thought possible. These types of classes might be provided in-house, but they may also be accessed online through distance learning from universities or via online-learning platforms (e.g. Coursera, edX, Lynda.com, Udacity, and so on, see Appendix 1). Never underestimate the value of a well-structured text to help you acquire ideas on how to best deliver your courses.

Consult a learning advisor. If you are researching in a university, the likelihood is that you will have a Learning and Teaching Center that employs support staff who are on hand to advise you on the best practice that you can take to promote learning in your student body. This advice might come in the guise of helping you to put together your unit, developing digital course delivery approaches, or outlining the expectations for assessment. This type of support can ultimately save you time in the establishment of your teaching units.

Promote self-directed learning in your students. The key to promoting better knowledge acquisition is to learn through practice and application, instead of the traditional tertiary approach of learning by wrote and adhering to hefty textbooks. Problem-based learning, student-centred learning, and investigative learning are modern styles of education at the tertiary level that can significantly impact your students' learning capacity.

Use your network. A solid approach to aid the effective development of your own courses is to consult established colleagues who are already delivering courses in your faculty. Developing a local network of peers can guide you and provide feedback where required.

BE INNOVATIVE IN YOUR TEACHING, AS FOR YOUR RESEARCH

Vary Your Assessment Styles

Academic assessment in undergraduate STEM disciplines typically takes the form of written essays, multiple choice or short answer questions, research practical reports, and group projects/problem-based learning. Education is constantly evolving, however, and you need to evolve with it. You have digital and electronic tools at your fingertips that you can employ, and novel approaches that you can take for assessing students.

Blog posts to gauge interpretation of a research article. This approach can gauge the student's comprehension of a research article, thus improving their capacity to extract key information to provide an overview of the key findings. In tandem, blog posts are typically short, easy to read and assess, and present an opportunity to encourage students to reflect on the material that they are required to learn.

Reflecting on data outcomes is required throughout a research career, thus this approach provides good training.

Embrace oral and video presenting as an assessment tool. As a researcher, you will be assessed on your capacity to write about your work, but you also need to present your work in a visual format and through digital means. Having your students complete presentation tasks as part of their courses can promote this skill development, whilst minimizing the time it takes you to assess the work. This contrasts with the extensive amount of time and concentration it takes to assess written assignment material.

Digital learning interfaces. Digital pedagogies are evolving in tertiary education: many institutions routinely deliver their units with online unit interfaces and support. This makes the delivery of courses more flexible for you as a lecturer, and provides the students with a simple interface from which to source resources and communicate with the teaching and tutoring teams. Innovative approaches to developing a digital course delivery are often supported by learning advisors employed by the university.

Departmental Responsibilities

When you take on a new departmental position, numerous extra responsibilities can be placed on you. These may or may not be directly relevant to your research discipline, but regardless will present a range of benefits and disadvantages. These will be discussed here.

Management of facilities

Some tenure-track posts can be secured through your capacity to develop and manage a resource or facility in a faculty or department. The management of this resource can secure not only your role in the institution, but it can also provide access for your group to an essential resource to promote your research progression.

Problems often come about through managing the users of the facility or resource for which you are responsible. Thus, well-structured guidelines should be established at the outset, in addition to regulations for users.

CASE STUDY: Facility management

A new group leader was recruited to a research institution to establish and lead a facility on site. This represented the first of its kind to be established in this country, and the group leader brought many years of expertise in running projects using this technology as a research fellow. It took an enormous amount of time and effort to establish the facility. In the end, the group leader was given a permanent technician to manage and run the service under their guidance. Technical staff were provided on-site to handle the logistics of the facility. The result was easy access to an essential resource otherwise would have had to be sourced from contractors, which would have required more time and investment. The success of the facility not only secured the group leader's position within the institution and provided bargaining power for pay and work conditions, but it also promoted extensive collaboration with other groups, both inside and outside the institution. This expanded publication output significantly and ensured ongoing funding for the group.

This case study demonstrates the power that the extra effort of managing a core facility can have over your capacity to establish your own niche. It is also associated with improved stability in your career and group.

MANAGING A CORE FACILITY CAN EXPAND YOUR PRODUCTIVITY

The management of an in-house core facility usually also requires that you develop and deliver training for users, in order to ensure that they operate effectively.

Coordination of postgraduates

Many university departments and faculties establish an in-house representative to manage issues relating to postgraduate research students, and to manage matters that may arise relating to their candidature. This role can bring numerous challenges, particularly when managing situations of conflict between the students and their supervisors (i.e. your colleagues), which can often devolve into significant drama. However, this role brings greater networks with other researchers within the institution, including the executive at the university. This can provide a platform upon which a research leader can set themselves apart for professional promotion, as well as making themselves more visible as a preferred mentor to future postgraduate students.

Organization of journal clubs or weekly seminars

As a group leader and tenured faculty member, you may be tasked with managing the weekly seminars or journal clubs for the department. Although this might be a logistical pain, it is a great opportunity to rally the involvement of postdocs and postgraduates. As such, if assigned this role, it can actually provide you an extra opportunity to mentor others in **developing management skills**. Organizing this type of regular departmental event is a solid way for young researchers to develop skills in **networking, organizing and chairing sessions, and managing a schedule of events**.

Management of annual events / open days

Many academic institutions will participate in open days to share with the general public the work that's being performed by academic teams. This represents a good opportunity for all group members to practice their **scientific outreach and general communication skills**.

Finding innovative ways to publicize your work to the public, whom usually indirectly fund your work with their tax dollars, will help in the public awareness of your work and contribute to an improved research profile for you. Similarly, even if these types of events end up with a poor turnout, they are solid opportunities for department/faculty-level team building; they are a chance to work together with people from other groups and sections, which you will need to do to make sure the events run smoothly.

Committees

Academic departments and faculties will generally have a number of committees to manage resources and other matters that form an integral part of the successful functioning of the institution. As a tenure-track member of the department/faculty, you will be expected to participate in various committees. These might include committees overseeing anything from postgraduate student affairs, important facilities and operations, research, learning and teaching, executive matters, and staff concerns. These are all very important to maintain communication and coordination between departments and faculties to ensure the smooth running of the core business of your institution, learning, and research. Like any good democracy, these fora also offer the opportunity for you to have a say in how your business/department/faculty operates. They give you the chance to ensure that your interests, and those of the parties you are representing/advocating for, are taken into consideration with all major management decisions.

> **COMMITTEES SUPPORT ADVOCACY FOR YOUR INTERESTS**

Recruitment

As an active member of your institution, you will probably be asked to participate in recruitment processes from time to time, whether the applicant is earmarked to work for you or not. This may involve sitting in on interview panels, reviewing applicants that have applied for roles in order to create a shortlist, or participating in discussion groups to select the best candidate. This experience may seem like a distraction from your other commitments, but aiding the selection of good candidates is a collegiate activity. Those colleagues that you assist will likely return the favor in the future.

Time Management of Duties

Regardless of the teaching or departmental commitments that you take on, effective time management is essential to incorporating these responsibilities into your busy schedule. When drafting your routine schedules (chapter 6), consider the time you will require each week to achieve the tasks involved in your roles. This can be refined to a routine protocol that you follow once you practice the anticipated schedule and adjust it so it fits well into your schedule. Improving your contribution to these types of institutional responsibilities requires the honing of this type of skillset, much like you honed your skillset to become a research academic in the first place. With effort, the associated benefits of your role will ultimately outweigh the time you have spent to contribute to it.

Table 8.1: Overview of Teaching Responsibilities

1: Teaching practical classes
MUST? Yes. Part of tenure role.
PROS:
• Potential to deliver practical classes related to own work.
• May identify talented potential future postgraduate students.
CONS:
• Time consuming.
CAN YOU DELEGATE?
• Postgraduates can act as teachers/tutors in the classes.
• Postdoctoral fellows may assist in class organization.
2: Teaching and coordinating courses
MUST? Yes. Part of tenure role.
PROS:
• Postdoctoral fellows may assist in class organization.
• Potential to deliver material and assignments related to own work.
• May identify talented potential future postgraduate students.
CONS:
• Time consuming.
• Dealing with issues.
CAN YOU DELEGATE?
• Postdoctoral fellows should be encouraged to teach some classes and learn how the courses are developed and delivered.
3: Mentoring/leading postgraduate research students
MUST? Usually. You may select candidates.
PROS:
• Expands research output.
• Provides potential collaborations through joint projects.
• Extra funding through student-specific grants and resources.
CONS:
• Potential conflicts may arise during candidature.
• Unexpected costs may arise which yield no equal outcome.
• Work may not yield expected outcomes.
CAN YOU DELEGATE?
• Joint supervisor roles are possible.

Table 8.2: Overview of Departmental Responsibilities

1: Managing facilities
MUST? Depending on your contract conditions.
PROS:
• Potential for extra staff under your command.
• Easy access and control over essential research tool.
• Easy collaboration with others requiring facility/resource.
CONS:
• Time consuming.
• Potential user conflicts.
CAN YOU DELEGATE?
• Possible technical staff can manage the day-to-day operations.
2: Coordinating postgraduates
MUST? No.
PROS:
• Expanded networks.
CONS:
• Required to manage conflicts.
CAN YOU DELEGATE?
• No.
3: Organizing journal clubs and seminar series
MUST? No.
PROS:
• Expanded networks.
• Training for your staff and students in literature reviewing (JC) and giving seminars (S).
CONS:
• Challenges in coordinating others — drop outs/ins, scheduling.
CAN YOU DELEGATE?
• Represents a good opportunity for postdocs and PhD students to hone their organizational and networking skills.
4: Annual events and open days
MUST? Yes.
PROS:
• Expanded networks.
• Honing science outreach and communication skills.
CONS:
• Time consuming.

- Can induce conflict.

CAN YOU DELEGATE?

- Your students and staff can work together with you to design your group's contribution.

5: Committees

MUST? No.

PROS:

- Expanded networks.
- Direct activities and resources to your interest.

CONS:

- Time consuming.

CAN YOU DELEGATE?

- Sometimes others can attend in your absence.

6: Recruitment

MUST? Yes/No.

PROS:

- Expanded networks.

CONS:

- Time consuming.

CAN YOU DELEGATE?

- No.

END OF CHAPTER 8 SUMMARY
Teaching and Departmental Responsibilities

In this chapter we have discussed how you might optimize your approaches to departmental and teaching commitments whilst maximizing the "value added" that you attain. We also discussed how you might include your research team in the activities related to teaching and departmental responsibilities. The key messages from this chapter include the following points.

- Structure your courses in a way that you might incorporate some of your research team into teaching and tutoring, providing professional experience for your team.

- Maximize the "value added" from your teaching by incorporating your current research interests into the course you are teaching.

- Pursue innovation in your pedagogical approaches by consulting educational designers, advancing your pedagogical training, and embracing novel assessment strategies.

- Actively consider the pros and cons derived from taking on departmental responsibilities before accepting or declining them.

REFLECT

1: What approach do you take to managing your teaching responsibilities? Do you share the teaching load in any way?

2: What kind of courses have proven the easiest for you to teach and why? What did you like the most/least about the experience?

3: What departmental responsibilities have you held, and have these benefited your work?

EXERCISE 8.1: VALUE ADDED

Engaging in teaching and departmental responsibilities adds value to your position as a tenure track academic for the institution with which you are affiliated. Take a moment to consider what you are contributing to, and the relative pros and cons regarding how it impacts you.

Make a list of teaching and research responsibilities in your department, and outline the relative responsibilities and estimated time commitment for each. Similarly, outline the potential benefits that you would get from contributing to this role. See table 8.1 for reference.

Activity	Must?	Pro	Con	Can you delegate?

EXERCISE 8.2: TEACHING

Make a summary of the last five courses to which you have significantly contributed because of your academic role. List the styles of teaching that you have used in your courses. Teaching style might include the mode of delivery (lecturing, online delivery, tutorials, practical classes, and so on), and class size. Note the types of assessment that you employed (blog posts, exam, assignment, oral presentation, or lab report). Then note the level of support you received from your team. Use this completed exercise to reflect on the successes and challenges that you encountered in your course delivery, and what you might do differently to improve on your future courses. This table can be downloaded from practicalacademic.com, along with an example of a completed exercise.

	Course	Teaching style	Assessment	Group support
1				
2				
3				
4				
5				

EXERCISE 8.3: GROUP SUPPORT

Make a list of your current team and their research foci. Try to create a mind map of the types of courses they might contribute to at the undergraduate level. Begin a discussion with your postdoctoral fellows and postgrads to gauge their interest in teaching or tutoring, whether paid or unpaid, and work together with them to identify where their work might benefit from teaching and tutoring commitments.

Chapter 8 — Downloadable Materials
Download from www.practicalacademic.com

- Table 8.1 and exercise 8.1: "Value added" — Excel file containing a blank template and the example provided in table 8.1.
- Exercise 8.2: "Teaching" — Excel file containing a blank template and an example of the completed exercise.

SECTION 1 SUMMARY
Eight Things You Can Do To Be A Better Group Leader

We have discussed a number of topics related to managing your research group effectively. Here we list the take-home messages from each of the eight chapters of Section 1.

1: Take responsibility for your group and be proactive in developing the business of your research group (chapter 1).

2: Structure your group to be productive, and do so in a way that suits your stage of group development, your style as a manager, and your discipline (chapter 2).

3: Choose your people well and encourage their individual professional development (chapter 3).

4: Proactively manage your budget and ensure safety in the group, both of which are essential components of a functional and active research group (chapter 4).

5: Network, and promote networking amongst your staff and students (chapter 5).

6: Manage the group and individual group members' time effectively (chapter 6).

7: Promote a positive and supportive workplace that is free of misconduct and conflict: proactively pursue a cohesive group (chapter 7).

8: Be proactive and innovative with teaching and departmental responsibilities, considering the potential side-benefits that may stem from these pursuits (chapter 8).

SECTION 2: MANAGING RESEARCH PROJECTS

In this section we focus on projects, which are a creative and active component of academic life. Working at the cutting edge of knowledge advancement means you have no specific blueprint for how to apply your core research skills. Nonetheless, ensuring that you design, develop, manage, and deliver your projects effectively and on time is important, and those issues are discussed here. This discussion is presented in the following chapter structure:

- **Chapter 9** covers project design and creating a clear project plan.
- **Chapter 10** discusses the type of staff you will include in the work, and how.
- **Chapter 11** introduces the main considerations for getting your project started.
- **Chapter 12** explores the practical issues you will typically confront when the project is underway.
- **Chapter 13** considers the best approaches to troubleshooting and dealing with change.
- **Chapter 14** presents issues relating to communicating and networking your work.
- **Chapter 15** reviews authorship and publication ethics.
- **Chapter 16** focuses on project completion and future directions.

For ease of reference throughout this section, the following definition will be applied:

"project lead" or **"project manager"** refers to the person managing the project (i.e. the person directing a specific project). This may be a postdoctoral fellow, a postgraduate student, a technician, or a group leader. Who directs a given individual project will depend on the project, group structure, and individual group dynamic.

CHAPTER 9

Designing Your Project

In this chapter we discuss the issues confronting you regarding project design, and how you can best establish a research plan that you will typically use as a basis for funding applications. Projects of variable lengths will be discussed, including one or many staff members, but the principal focus here is on the classical 3–5 year project, which represents a standard length of funding in academic research.

Similarly, in this chapter we will discuss the involvement of collaborators, the establishment of brand new or ongoing lines of investigation, and the design of logistics surrounding your investigation.

Deciding on the Right Work to Pursue

It is absolutely critical when you design your project that you evaluate whether it will be something new to the state of the art. Regardless of where the project idea came from, you must take some time to research the literature and the current key players in your field, to determine if the idea has been reported previously and to gauge whether others may be pursuing the same type of work. You need to be as sure as possible that you are pursuing a **unique line of investigation** that you are capable of completing, and that all of the required elements are in place.

Nonetheless, it is entirely possible that other groups will be pursuing the same thing that you are working on in tandem. The more research you do to understand the key players in the field, and to network with them if possible, the better prepared you are to pursue your focal research interest. It is useful to have a **positive international network in your field**, so as to operate in a collegiate manner, rather than a competitive one. This is something that typically comes with time and experience in your field.

Your project may represent the next stage of a long-term research investigation in your group, or a completely new investigation based entirely on your review of the literature. In any case, the **project should be planned out systematically** with care taken to observe the fine detail, and it is essential that you anticipate potential setbacks from the very beginning. You should establish a solid budget, particularly if you are writing your project plan for a funding application, with contingencies in place for projected setbacks: always prepare for the worst-case scenario.

Finally, you need to ensure that you plan for the contributors required for the work, including the number of **participants: contractors, collaborators, staff, and students**. Think about who will contribute what to the work, how the labor will be distributed, and who will retain the rights over the work should it be ongoing once this project is complete.

SMART Planning

When designing a project, it is wise to consider how you will gauge its success. The measures of project success should be clear to the project team, and to those funding the work. The SMART method is a popular approach for determining what success means to a project (Doran, 1981).

Specific
Measurable
Attainable
Realistic
Timely

The SMART criteria can be applied to the academic research project to frame the value of the project that you are designing. The five main steps are outlined here in order.

Specific — you need to incorporate specific goals into your project, outlining achievable deliverables.

Measurable — determine clear, measurable waypoints and outcomes that represent goal attainment.

Attainable — make a clear outline of how the defined goals and deliverables can be practically achieved.

Realistic — ensure the goals you are working toward are those you are willing and able to pursue.

Timely — the goals you are aiming for must be achievable within the timeframe that you have outlined. Clear timelines must be set.

Taking the SMART criteria into consideration when you are preparing a project plan is helpful to ensure that you are designing your project well.

FOLLOW SMART GUIDELINES FOR YOUR PROJECT

Types of Research Projects

This section will discuss the issues surrounding project design and implementation. It will have no specific distinction regarding quantitative or qualitative research plans, because both approaches may benefit from the guidelines noted here. Most of the research in mind when developing this guide reflects on experimental research, which often incorporates blinding/double blinding, randomly-assigning subjects to treatment groups, and the testing of independent variables which are then manipulated.

Approaches to Project Design

Numerous styles may be employed when designing overall project plans, which will depend to some extent on the leader's discipline style and the project's individual goals. It is worth considering which style would best suit you.

Several classical project design approaches are listed here for your consideration. These styles are only representative, but they indicate basic principles that are often followed by quantitative researchers in academic institutions.

One-path approach

A simple project outline might be applicable for small projects for junior research students. This approach is typically represented by a project that is very achievable, and will see the researcher pursuing the work to reach achievable goals within the timeframe they are allocated. It is often a short and fast research investigation focused on obtaining a degree, certificate, or qualification. It may represent a focus on tidying up loose ends on a larger project that has been underway for some time, or investigating something extra to complement a larger research work. Sometimes, this type of work may represent a short preliminary investigation designed to explore a potential new research direction.

Several-path approach

In this approach, the research is usually focused around a major high-risk project that could yield high returns. The nature of the beast is that high-risk research projects often hit brick walls that can sometimes be insurmountable. Similarly, competing groups might achieve the same outcomes prior to your own group.

It is highly recommended to design low-risk "bread and butter" research investigations to perform in tandem, which are likely to yield likely low-impact but reliable research outputs/journal articles.

For example, a researcher might pursue small *in vitro* investigations whilst larger *in vivo* projects are in development.

Collaborative or independent / outsourcing
Projects can often benefit from collaborative approaches and outsourcing to facilitate project development. This may omit the requirement to set up expensive services and experimental approaches that could easily be achieved via petitioned services, which can be arranged through collaboration or outsourcing to commercial providers. Some examples of outsourced services may include: synthetic protein creation, antibody development, transgenic mouse injection and founder development, or clinical data retrieval and analysis.

CASE STUDY: Several-path approach
A junior researcher always pursued a four-path approach. This incorporates one high-risk/high-return project that can run in tandem with shorter, lower impact projects that can keep producing publications. This approach yielded an average of 2–3 low-impact papers per year and a high-impact paper every 2–3 years. Over 6 years, one high-risk project was outpublished 18 months into the work, but the researcher maintained their active research-publication output because the low-risk projects were underway.

Designing Stages
Each project path will follow stages in its development. These stages will vary for each project path being pursued. The stages are usually marked by micro-deliverables that represent waypoints on the path toward the overall project goal (Glen, 2009). Micro-deliverables may represent different stages in the progression of a larger project, or small steps towards achieving these stages. They might represent points of

achievement that should be reached within a weekly or monthly plan. An example is shown here.

PROJECT ONE — example stages
Stage 1: extract DNA clone to plasmid stock.
Stage 2: modify to introduce mutation of interest to cDNA.
Stage 3: create expression construct from mutated DNA.
Stage 4: create stable cell lines expressing either mutant or wild-type expression construct.
Stage 5: analyze changes in cell behavior in both cell lines.

This is an example of a project stage outline that might be pursued in cell biology, and which might represent the structure put forward in a funding application or student project. Creating this type of broad overview that is marked by defined achievements creates a framework for designing the more detailed components of the work.

Creating a Project Plan

Projects may or may not stem directly from grants that outline a detailed research plan. In any case, by the time the grant is awarded, chances are that there have been substantial changes to the state of the art, and these changes must be considered in the fine-tuning of the work to be performed. Regardless, it is wise to allocate time at the beginning of any research project to ensure that you are up-to-date with the most recent advances in the field, and to ensure that you are still focusing on completing work that remains unpublished. This is a good time to confirm your project goals, expected outcomes, contributors, collaborators, timeline, and milestones that you need to achieve.

GANNT charts
GANNT charts are incredibly useful for structuring your project plan in a visual calendar that outlines all of your major directions over the entire project period. A typical GANNT outline for a postdoctoral

two-year research project is outlined in table 6.1. GANNT charts can be applied to any type of project and are a useful addition to formal project applications. Such a chart allows those reviewing your funding application to gauge whether you have thought through your overall project plan and the deliverables that you will achieve throughout the work.

Similarly, a tabulated chart like this is useful to keep as a visual aid for tracking overall project progress, and for gauging whether the timing that you have outlined is achievable. They can be developed at the commencement of a project, and they can be modified as change takes place throughout the life of the project. An example GANNT chart for a three year project is shown on the next page.

Making a Project Outline

Making a project outline is critical to setting out your expectations at the creative development stage of the work. Once you have identified your reasons —preferably based on literature review and preliminary data — leading to the project in question, the timeframe, and the associated desired goals, you should develop a **1–2 page project outline** that provides a systematic and practical research plan which mirrors the timeline noted in the GANNT chart.

The project outline should clarify the reasoning behind the project, the main research activities that will address the proposed hypothesis, and the expected findings from these activities. It may include information about the people participating in the work, particularly if it is a large and collaborative venture. It may also do this if the work relies on the expertise or resources associated with a major contributor or service. A project outline is usually written as a summary in clear and flowing language, and it explains the progression of the work.

Table 9.1: GANNT chart for three year project

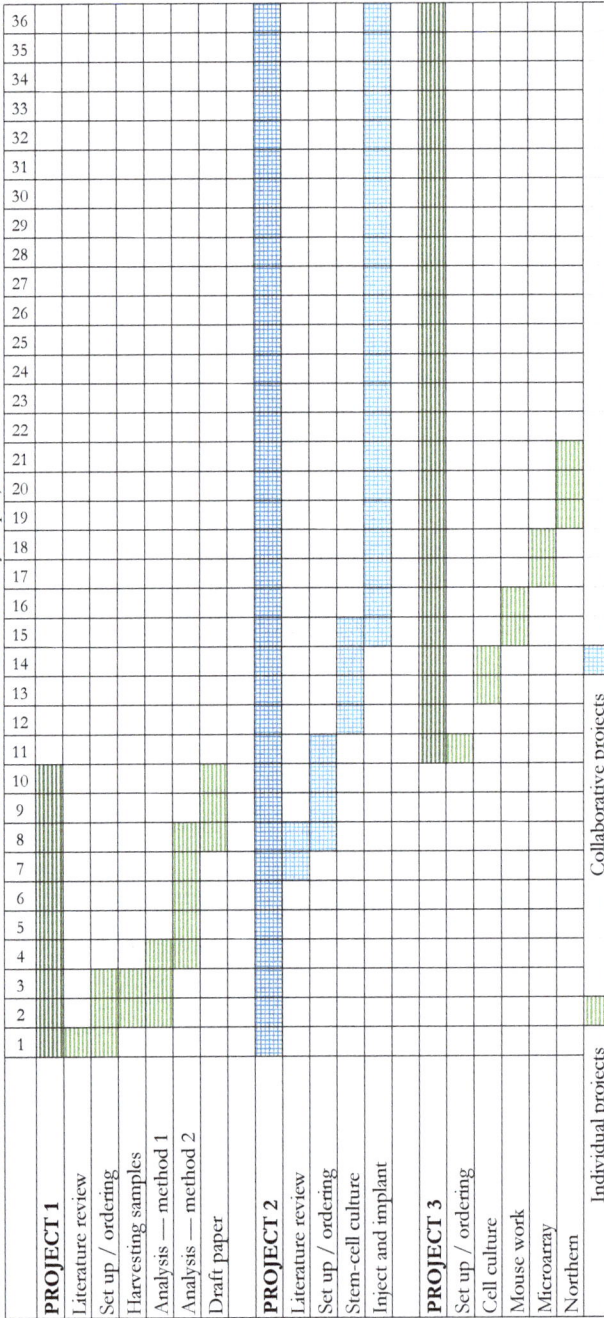

Task	Months (1–36)
PROJECT 1	
Literature review	1–2
Set up / ordering	2–4
Harvesting samples	4
Analysis — method 1	7–8
Analysis — method 2	8–10
Draft paper	9–10
PROJECT 2	
Literature review	1–2
Set up / ordering	8–12
Stem-cell culture	12–15
Inject and implant	11–36
PROJECT 3	
Set up / ordering	11–36
Cell culture	14–16
Mouse work	16–18
Microarray	18–21
Northern	
Individual projects	2
Collaborative projects	14

188

THE PROJECT PROPOSAL SERVES AS A BLUEPRINT

Developing a Full Project Proposal

The full project proposal will include all of the extra support data, detailed information regarding practical approaches, and information about the people who will be carrying out the work. It should provide as many links as possible to practical information regarding project approaches. This will enable it to act as a blueprint for project participants to refer to throughout their research endeavors.

Figure 9.2 demonstrates the components of developing a project proposal. The project proposal should provide a overview with detailed information that outlines the starting point from which the project can commence. Once the project is underway, this proposal should be used as a modifiable template, which can be updated with progress and changes in the state of the art that might impact progress or change the focus of the work being pursued. Given the complex nature of most investigative research projects, it is very likely that the scope and detail of the project will shift during the investigation.

Figure 9.2: Flow of developing a full project proposal

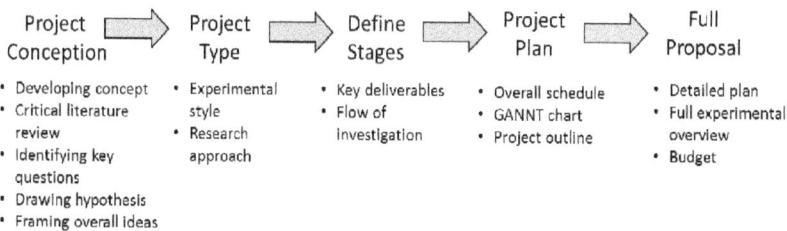

Project Conception	Project Type	Define Stages	Project Plan	Full Proposal
• Developing concept • Critical literature review • Identifying key questions • Drawing hypothesis • Framing overall ideas	• Experimental style • Research approach	• Key deliverables • Flow of investigation	• Overall schedule • GANNT chart • Project outline	• Detailed plan • Full experimental overview • Budget

This overview demonstrates the progressive development of a full project proposal from conception to a workable blueprint, from which the project can then commence. This also represents how project ideas typically develop for grant applications in order to fund the work. During the establishment of a project proposal, **preliminary data** usually plays a key role in defining the concept to be pursued.

Components of Project Proposal

Here we describe the key components and contents of standard project proposals. This outline is by no means exhaustive, but it is intended as a guide for project design and development.

Scientific abstract

The scientific abstract is written with the express intention of communicating an overview of your project to your professional peers. This style of abstract usually includes more detail regarding the project components, experimental approaches, and discipline-specific terminology, so that the scientific reviewer can gauge exactly the approach that you are using for the project.

Layman's abstract

The layman's abstract is intended for the non-scientific audience, so that they can gauge the relevance of the work, the overall strategies you are taking, and the key advances that you will (hopefully!) achieve. This abstract is written such that it may be utilized to publicize your research to a wider audience, and so that it may be included to general research communications on institutional or funding body websites, or in annual reports.

State of the art / literature review

The state-of-the-art/literature review provides the opportunity for you to frame the context in which the project is being pursued, relative to what has been achieved in the field to date. The literature review should clarify the gaps in the field that are addressed by your project; it should do this by highlighting areas that require attention throughout the review. This should culminate in the overall research question/hypothesis that you are addressing in your project, and in an argument for why it is critical to pursue and fund this work.

Preliminary data

Preliminary data may be introduced in the argument for funding the proposal, and may prove decisive for funding panels when they are gauging whether you and your team will be able to advance the work. Demonstrating that you already have established protocols, or showing key initial findings for the project that have been derived by your team, can represent the key to defining your capacity to undertake and deliver on the project.

Experimental plan

The experimental plan is the heart of any project proposal. It outlines the blueprint for how the work will be practically undertaken to address the focal hypothesis. This plan should break the project down into achievable stages, defining the key deliverables for each stage and the approaches that will be taken in the process. The experimental plan should define the **major goals, major resources, and projected challenges or delays** that may be encountered; it should also include contingency plans and ways that you could circumvent these issues, or alternative approaches that may be taken in case of specific project failures. If the project is divided into sections, the methodology should also be broken down into sections reflecting those outlined in the GANNT chart.

 GANNT chart

A GANNT chart (table 9.1) should be included in a project proposal. It provides a clear outline of the timeline for achieving all of the key project deliverables within the scope of the contract, fellowship, or grant that is driving the work. GANNT charts are discussed earlier in this chapter in the section entitled "Creating a Project Plan."

References

The inclusion of key references is standard in the state-of-the-art/literature review, but may also be used in the experimental plan to provide background for key project approaches. References are important for both project reviewers and — potentially — key project staff to be able to develop a deeper understanding of the project background and the approaches being taken. Here you should clarify critical studies that led to the conception of this work. When you use the project proposal to apply for funding, in particular, you should consider the strategic inclusion of published work by your group to frame the current project plan. Inclusion of your own published works shows that you are continuing on an established research focus for which you have previously delivered publications.

Key participants

It is critical to include key participants in a project proposal to ensure its funding success. Aligning the skillsets and backgrounds of those participating in the project to the actual work being proposed demonstrates the team's capacity to achieve the experimental goals. If the work you are outlining is cutting edge and promises to advance the field dramatically, but lacks participants with relevant experience in the key experimental approach, the proposal will be scored poorly. Thus, the inclusion of participants' CVs, outlining essential skillsets and a track record of achieving project goals, is advisable.

Budgeting a Project

When managing your individual projects and overall group funding (covered in detail in chapter 4), it is important to strategize your project funding. When it comes to establishing project budgets, **overall expenses that are expected for the entire project should be outlined** in detail before commencing. This should cover all projected input expenses, labor expenses, and equipment/activity expenses.

Effective grant budgets provide realistic expenditure plans that support project advancement. Such advancement involves performing work that will yield a relevant outcome, be it advancement of knowledge, striving toward developing a useful tool, or finding a cure. Established project participants who already have scholarships or fellowships strengthen the project, making it more attractive to funding bodies because they can omit the cost of an active researcher when allocating budgets for funding.

Ethics Clearances and Safety

Gone are the days when you could start research work that does not require all of the i's to be dotted and t's to be crossed. Ethics clearances are essential for any work, although they are particularly essential for working with dangerous substances, patient samples, potentially dangerous pathogens, or animals. **Ethics clearances must be acquired well in advance of the project commencing**. This is the responsibility of both the group leader and the project manager (if they are two different people).

Overall group and departmental safety issues should be managed by institutional staff and group leaders, but project-specific safety issues often require safety clearances in order to commence work outlined for projects. Examples of this include working with dangerous restricted chemicals or materials, accessing endangered areas for sample collection, and using animals (particularly in research investigations). Thus the project manager needs to ensure that all appropriate safety precautions are observed throughout the project's lifespan. This issue will also be referred to in later chapters. At the project design stage, you and/or the project manager should develop a summary of expected ethical and safety considerations, and incorporate it into the overall project proposal.

Dealing with Negative Reviews on Grant Applications

Although applying for grants is not a major focus of this book, it is worth mentioning the issue of being knocked back on grant applications you make with your project proposal, and how you might react/respond to this. Any number of reasons may have led to the rejection of your funding application, but it is a growth experience, and it is important to benefit and learn from the experience.

It is essential to reflect on the weaknesses of your application and focus on what you can do to improve in the next funding round.

The funding body may provide you with specific details of why your application failed to make the cut. Your project may have been overlooked due to weaknesses in design, practicability, or relevance related to the focal interests of the call. The track record of the participants whom you included to the application may be too weak, and thus uncompetitive, or their relative experience may not be appropriately relevant to the proposed work.

The takeaway from this is that you should not throw in the towel on a given project when it is rejected for funding. **Redesign, strengthen, gain more preliminary data**, expand the participants to include collaborators, cast a wider net for funding from more sources, and continue to build your argument for why the project should be pursued.

Improving Your Project Proposals

Once you have developed your full project proposal, it is important to then review the overall document. Several approaches can be taken to do this.

Self-appraisal — assessing your own project proposal from a critical viewpoint is essential to strengthening your design. Only through several rounds of revision will you yield a strong proposal that will deliver solid research outcomes.

Team-appraisal — having your team review the overall proposal can dramatically strengthen its design. It is especially important you get the team members who will work on the planned project to review the proposal. Your team has a vested interest in the topic, and they may bring a more practical perspective to the projected approaches, timeframe, and achievability of the outlined work. Similarly, having those researchers who are more junior review the project proposal helps to train them in learning to write project proposals themselves. This reflects an apprenticeship-style of training in all aspects of the profession.

<u>Peer appraisal</u>

If you are an established researcher, chances are that you will have a peer network, not only from your historical professional experience, but also in your current institution. It is worth networking specifically with those not in your immediate research field to assist each other with reviewing one another's project plans, in order to cast a critical eye and strengthen the design. After all, these are the types of people that will ultimately be reviewing the funding applications you are applying for, and also potentially reviewing the manuscripts seeking publication stemming from the work. Some departments establish informal reviewing sessions to encourage their researchers to participate in helping each other to develop their proposals prior to major grant rounds.

Learning How to Project Manage

Reading this book should give you a good start in figuring out how to manage your projects and/or groups, but you should never stop trying to improve upon your work. If you wish to pursue a life in academic research, you will take on increasingly managerial roles, and thus skills in this regard can only aid your professional growth. Managing others effectively, to achieve professional satisfaction in your staff and thus promote productivity, is key to your own successful professional growth and fulfilment. If you are setting up to manage a project or group of people who are running projects under your supervision, **it pays to proactively learn the art of project management**. Investigate what options you have for online learning or in-house courses run at your institution. You may spend some time learning what you think is common sense, but you will also pick up some key approaches that can revolutionize the effectiveness of your leadership.

END OF CHAPTER 9 SUMMARY
Designing your project

In this chapter we have discussed how you establish and develop your project. This covered project planning to develop a fully-functional project outline, including budget, timeframe, clearances, and other concerns. The main take-home considerations for project design included the following.

- Choose a project focus that develops a unique line of investigation, and then plan systematically, selecting solid contributors to the work.

- Follow SMART criteria to determine what success means to your project.

- Consider the best design/approach that will benefit your project, incorporating stages to your project plans so you can track the progress of the work.

- Develop a GANNT chart to better review the overall project plan.

- Create a project outline to create a short overview of the overall plan.

- When developing a full project proposal, ensure all of the required components are incorporated, including a clear budget and a clear outline regarding ethics clearances and safety considerations.

- Make the most of feedback on negative project proposal reviews, and utilize your networks to help build a stronger project for another round

- Embrace opportunities to develop your own project management skills and those of your project managers, prior to the commencement of the project.

REFLECT

1: Do you take into consideration all aspects of the project and plan effectively before beginning major new investigations?

2: Have your project proposals effectively won funding in the past? What are you doing right? What could you do better?

3: Do you budget effectively and ensure that you set up proper ethics clearances and safety guidelines for all of the work underway in your group?

EXERCISE 9.1: PROJECT PLANNING

What kind of style do you use when designing your projects? Consider the approaches to project design that have been talked about, and draft a list of all of your successful grant awards. Identify the style that you have applied in your project design. Did it work well, or could a different approach potentially result in improved outcomes for your work? Did you employ a one-path, several-path, collaborative, or independent/outsourcing approach?

Grant	Project style	Reflection

EXERCISE 9.2: GANNT CHART FOR PROJECT (STAGES)

Create a framework for a project you have in mind by identifying the waypoints or micro-deliverables that you anticipate achieving in the scope of the work. Consider linking multiple project stages together to create a complete three-year project plan with multiple streams and potential deliverable outcomes, such as publication or patentable technologies. Consider creating a GANNT chart to demonstrate this graphically. A template can be downloaded from the Practical Academic website (www.practicalacademic.com).

EXERCISE 9.3: REVIEWING GRANTS/PROJECTS

Find a colleague in a similar but non-competing field and share your most recent grant applications or project plans with each other. Act as a non-biased reviewer of the work. What do you think your colleague could do better to improve their grant application or project planning success rates? This exercise will provide you with the opportunity to apply your understanding of project design with reflection on the concepts described in this chapter.

Chapter 9 — Downloadable Materials
Download from www.practicalacademic.com

- Exercise 9.1: "Project Planning" — Excel file containing a blank template and a completed example for reference.

- Exercise 9.2: "GANNT chart for project (Stages)" — Excel file containing a blank template and an example, as provided in table 9.1.

REFERENCES – Chapter 9

Doran, G. T. (1981). "There's a S.M.A.R.T. way to write management's goals and objectives". *Management Review*. AMA FORUM. **70** (11): 35–36.

Glen P (2009) Monitor project progress by using micro-deliverables. Tech Decision Maker, accessed 5/9/2016. (http://www.techrepublic.com/blog/tech-decision-maker/monitor-project-progress-by-using-micro-deliverables/)

Department for Business Enterprise and Regulatory Reform, UK, Guidelines for Managing Projects, August 2007 http://www.berr.gov.uk/files/file40647.pdf

Creswell JW. 1998 Qualitative inquiry and research design. Choosing among five traditions. SAGE publications.

CHAPTER 10

Strategic Planning — Project Staff

How many people do you need to put on to a project, and is the project suitable for a given person? Matching your project to your students and staff is essential if you want to maximize the output from the project you have outlined. It is advisable to take the needs of the individual staff member into account, and to match their professional focus to their relative project contributions.

Internal Contributors to a Project

Postdoctoral fellow

The postdoc will often be instrumental in guiding the direction of their own work, and they should be encouraged to create and plan out their own projects. For junior fellows in particular, and for new fellows joining your group, you should provide support in project development. Junior fellows may be assigned a project to which they will be encouraged to contribute planning, but on which they also receive clear guidance.

PhD student

It should be assumed in the majority of cases that PhD students have little or no experience in research project development, and a project outline is often already in place when the student commences their doctoral studies in your group.

PhD students should be encouraged to work towards project goals that are achievable within the scope of their project timeframe, of which only 60-80% may actually constitute experimental work. Time allowances should be made for planning and starting up the project at the beginning, as well as time to write up at the end of the work.

The mentor/group leader should have a clear idea of which projects they would like the student to pursue at the commencement of their candidature, and they should encourage the students to develop a full research plan in their first two months in the role. PhD students should expand their project leadership skills as they progress through their training, like an apprentice, to be able to manage independently by the end of their candidature.

Masters or Honors students
Similar to PhD students, Masters or Honors students will typically lack experience in research project development and management, but they usually desire the development of these skills during their training period in your group. These students seek the completion of a simple one-year project. This project should actually encompass about 6–8 months of experimental work, depending on the institution. The experience should prepare the student to work as a support of experimental research in a technical role, or to continue on to pursue a PhD program.

These students should be assigned a low risk project that will provide some solid research outcomes within the limited timeframe of their degree program. This may be a part of a project from a larger funded program, or the establishment of preliminary data that might lead to funding for a more in-depth project that the student might take on as a PhD student.

Technicians
Technicians can participate in any project, and they carry no requirements with regard to accomplishments from a specific project

focus; thus they can be included into any project that is planned or currently underway in a group. Nonetheless, technicians are typically hired to academic groups on contracts mirroring funding that has been awarded for an established project plan.

Summer students

Summer students join research groups to gain professional experience in a real research scenario, creating reports of their findings from the training period. Their supervision can create an extra burden on postdocs and PhD students, who are typically required to manage their practical supervision, but hosting summer students entering their final undergraduate year can be a good way to find potential future postgraduates. Projects will typically shadow those already underway in the group, and their supervision provides a good opportunity for more experienced group members to gain experience as mentors and supervisors.

Table 10.1: Project types for different professional roles

Role	Experimental duration	Type of role	Number and type of projects
Postdoctoral fellow	2–3 years	Project lead/manager	Multiple — high and low risk
PhD student	18–30 months	Mentored researcher and/or lead	Multiple — high and low risk
Master's/ Honors student	8–9 months	Mentored researcher	One — low risk
Research technician	2–3 years	Support	Various/where needed
Summer student	5–7 weeks	Mentored researcher	One short-term, support work — low risk

Project teams

Postgraduate students and technicians usually require overall guidance on the progression of their projects. The creation of project teams, as

202

outlined in the team approach (chapter 2), can facilitate improved overall project productivity. This type of approach can maximize productivity on the work and allow a group to approach more than one project in tandem, to reduce the risk of individual failure.

Tailoring projects to your people
A quick summary of project types is noted in table 10.1. Overall, project type should be associated with the relative brevity of the role, and the requirements of the person pursuing the project. In a best-case scenario, by the time you are established in your own research group you will have numerous project ideas and several active research foci underway at any one time. As the group leader, you should be overseeing the broad focus of your group, with a clear idea of how all of the projects underway connect to each other and how they relate to your broader academic field. You should be able to gauge whether a project is high or low risk, high or low impact, and roughly how long it will take to complete.

For the more long-term research roles, in terms of academic research project contracts, **postdocs and PhD students are the best choices for taking on slightly higher risk projects**. They should be able to manage several projects in tandem, depending on the nature of the work, and thus should be able to explore new avenues, or take on higher-risk or longer-term projects, whilst maintaining the "nuts and bolts" basic work that will ensure that they get papers out for publication. This is even more effective when they are working together as a team.

Honors, Masters, and summer students should be assigned less risky, shorter-term projects. These might be similar to the smaller projects that keep papers coming out for postdocs and PhD students, or perhaps those projects which contribute to a larger project for the more senior lab members. From the point of view of the student, they require an easy "taster" project that will give them some transformable data that they can relate back to the state of the art and create an assessable report or thesis on. If the short project holds potential to expand into a longer-term PhD project, all the better to lure the student into the group. From the point of view of the group leader, the training of these junior group members does take time and energy away from more directly rewarding pursuits in the short term. Training these students can result in a longer-term professional relationship, however, that can yield significant dividends if the student decides to stay on in the group.

Technicians and research assistants are usually employed off your own research grants or as part of an institutional agreement to provide assistance for, as an example, the management of a core facility. These staff members often stay in the group even longer than people in any other role, and they can take on any number of projects and tasks required in the group, from managing a facility to establishing a project. They are generally not expected to deliver publishable manuscripts, but rather they are expected to develop the core data to contribute to a paper. There are no hard and fast rules regarding the specific projects a technician is required to contribute to, but effort should be made to ensure that they actively participate in the project for which their funding stipulates their employment.

CASE STUDY: Project assignment

An established group was operating according to the team approach (chapter 2), in which postdocs, PhD students, and technicians worked together to develop multiple projects. In order to define the projects that were assigned to each team member, the team would divide up the work according to the team members' requirements, relative time, and project requirements. In this manner, each team member was able to maximize their productivity and recognition, and so were able to deliver on their professional requirements.

External Contributors to a Project

<u>Joint leadership of project</u>

Supervising research students or staff together with other group leaders can often expand your research interests significantly; it is, effectively, collaborating via the joint supervision of a researcher pursuing work at the interface between two groups. This works very well for projects that span two disciplines or research specialties, and usually positively impacts both groups and the researchers participating. It provides an opportunity for two areas of expertise to innovate in a collaborative fashion through the key participant: the researcher. Care should be taken to ensure clear communication and regular meetings between the student and supervisors throughout the project, because students can end up with minimum supervision from both parties or one-sided supervision that impedes the overall progress of their work.

CASE STUDY: Joint leadership/supervision

A PhD student commenced in a research group focusing on the role of a signaling protein, but the group leader wanted to expand the *in vitro* approach to create an *in vivo* model. Due to the PhD student's lack of experience in working with *in vivo* models, a joint supervision approach was taken together with another group leader in the department. That group leader had recently established the relevant facility on site.

During the course of the candidature, the student worked between the two groups, conducting the *in vitro* work under the guidance of the original supervisor and the *in vivo* work together with the associate supervisor and his technical staff. At the end of the candidature, the student succeeded in establishing a new *in vivo* area of investigation for both groups, publishing numerous manuscripts, and had two senior referees for future fellowship/postdoctoral applications.

Outsourcing

Facilities or resources may sometimes be challenging to acquire, or a particular part of the work may not be cost effective to set up. Thus outsourcing through contractors or established support services can often speed up the progression of your project, leaving your in-house staff and students more time to work on other things.

OUTSOURCING CAN MAXIMIZE PRODUCTIVITY OVER TIME

Collaboration and shared resources

CASE STUDY: Collaboration

A postgraduate student was brought in to develop a long project working between two supervisors in the establishment and testing of a new mutant zebrafish model. The student gained support from both supervisors' areas of expertise in the development of this model and ultimately delivered several high impact papers for both groups in the investigation. The model developed went on to form the basis for numerous investigations in both groups moving forward, and the collaboration persisted over many years and several collaborative research students and staff.

Collaborations provide an alternative to outsourcing. They can get essential work done in exchange for acknowledgement on a manuscript. Sharing resources can also facilitate project advancement.

Loss of IP or research focus is a potential risk due to the chance that the collaborator could take the resources and expand their research moving forward in competition with your own group, so collaborations should be entered into cautiously. Similarly, the collaborator could take over a research pursuit, essentially making your research their own domain. Clear guidelines regarding ownership of the work should be established and controlled for at the outset of any new collaborative research pursuit.

Visiting junior researcher

A junior researcher may arrange to work temporarily in a host group in order to complete part of their work in a location where an essential resource is accessible. This might pertain to a facility or resource available within the host group (like transgenic animals, microscope facilities, or established cell-culture set-up), or a facility that hosts many researchers (such as The Large Hadron Collider, a research field post, or a telescope). These visits might span weeks or up to a year. When joining a host group, it is typically understood that host researchers assisting the project will receive credit for any assistance they provide during the visiting researcher's time with the host group.

CASE STUDY: Visiting junior researcher

A junior researcher developed a test system to measure a key molecule in blood samples. All of the locally accessible samples were already tested. In order to validate the test system, and test it on a broader array of test samples, a collaboration was arranged with a foreign group that had access to an extensive collection of clinical samples. The junior researcher travelled to the host collaborator's group and worked for three months, setting up the test system and completing the extended analysis. This resulted in a stronger publication outcome and ongoing collaboration that persisted for some time.

Visiting sabbatical researcher

After establishing a productive research group and several years in a tenured position, senior researchers can take a sabbatical in other institutions, usually internationally, in order to refresh their perspectives and research interests in a stimulating environment. These typically last for approximately 6–12 months. They may agree with the host group leader that they will explore new research areas of interest in the host group, or they may actively pursue collaborative idea-generation with researchers at the institution. A sabbatical presents the possibility to expand perspectives in a fresh environment, surrounded by new people with different backgrounds to the researcher's home institution.

TABLE 10.2: Project types for different external project contributors

Role	Experimental duration	Type of role	Number and types of projects
Joint project leader	2–3 years	Project lead/manager	Multiple — high and low risk
Contractors/ service agents	Various	Service provider	Various
Collaborator	Various	Acknowledged contributor/co-author	Typically one-project collaboration
Visiting junior researcher	Typically weeks–one year	Mentored researcher and/or lead	Typically one project
Visiting sabbatical researcher	Typically 6–12 months	Independent investigator	Various

Inclusion of Contributors to Project Plan

You should consider including project contributors when you make a project plan, particularly in order to apply for grant money. The specific focus of making applications for project funding is beyond the scope of this book; however, the inclusion of participants with a strong curriculum and experience in the field for which the funding application is submitted can make the difference between a successful funding bid and rejection, regardless of the strength of the project plan.

The key points to consider when including contributors to a funding-application-focused project plan include:

- Overall professional excellence according to CV.
- Experience performing the techniques that are described in the project approach.
- Success in completing a project in the field previously, which may be shown by:
  - papers published;
  - patents;
  - theses delivered; and
  - technical support provided in the field.
- Being an established provider of a resource or technology that is essential for project completion.
- Availability to participate in the project:
  - project contributors or staff must demonstrate that they have sufficient time available to deliver on the project being proposed.

END OF CHAPTER 10 SUMMARY
Strategic Planning — Project Staff

In this chapter we have discussed the most appropriate way to staff projects depending on project characteristics. We compare the role that internal staff may play in projects of various risks and durations, and what kind of contribution might be made by participants external to your institution. This chapter covers the following key concerns:

- The inclusion of staff to projects should be strategically planned.

- Internal project staff should be matched to the type of project designed, in order to best align their capacity to requirements.

- Contributors external to the research group can strengthen the project capacity, and this can help to maximize on time and cost efficiency.

- The relative expectations of any project contributors, both internal and external, should be clarified at the outset.

REFLECT

1: Have you engaged the appropriate research staff for your projects in the past — that is, people who have demonstrated success in achieving their research goals?

2: Do you engage external contributors for projects and, if so, do you make clear agreements prior to entering into the work with them?

3: Do you strategically consider who you include in your project plans when applying for funding?

EXERCISE 10.1: STAFFING PROJECTS

Consider the last few projects that you have had running in your group and complete the following table, which is modeled on table 10.1. Record the staff member, their role in the context of the project, the duration of the project, the risk of and approach to the project, and the outcome that you observed. Was it successful; did that person succeed in the role?

Role	Type of role	Project duration	Project type	Outcome (+/-)

Chapter 10 — Downloadable Materials
Download from www.practicalacademic.com

- Exercise 10.1: "Staffing Projects" — Excel file containing a blank template and an example.

CHAPTER 11

Getting Your Project Started

The project has been designed and funding has been successfully sourced. Now you need to get the project off the ground and running. Where to begin, and what needs to be done? Here is an overview of major concerns that you should address whilst you are setting up.

Installing a Project Manager

Once the funding is confirmed and the project can commence, you need to get it up and running. The first thing to do is to **confirm the project manager/lead (PM)** who will be directing the research. This is typically a PhD student or a postdoc.

The group leader needs to establish the contract for the project lead, if not already in place, and meet directly with them to establish the overall project plan, to set initial tasks and expectations for deliverables throughout the course of the project, and outline the rest of the project team who will be included to participate in the work. The relationship between the group leader and project manager is critical to overall successful development and completion of the project; thus, this relationship should be defined at the outset.

DEFINING THE PROJECT MANAGER IS THE KEY FIRST STEP

The Project Manager's Start-up Tasks

Update the project

The project manager conducts an **update of the project plan and literature review**, identifying any changes that have taken place since the project was initially designed. This should be done systematically throughout the project, because it is critical to be completely up-to-date with the state of the art before initiating the project. Evaluating the current status of all project concerns is an essential first step.

Create a project schedule

In developing the practical project plan, **a GANNT chart** should be set up for the entire project over the duration of the project/funding. The chart should outline project goals and deliverables, and should be modified from the original funding application if required. The chart should include all reporting dates to the funding body and projected completion dates of research goals. See chapter 9, table 9.1, for an overview of project GANNT charts.

Participant outline

Definition of the project participants and their roles should be established at the beginning of the project. Specific project goals should be defined for each of the project participants. These goals should be outlined prior to the commencement of the participant's role in the project. Whether the project is being managed and pursued alone by one individual or not, the projected involvement of external project participants — including consultants, outsourced service agents, or collaborators — should be defined at the outset.

Permissions and clearances

Ethics clearances for all stages of the project need to be applied for and acquired if not already done so, to ensure that there is no delay in project progression. All members of the team should be cleared prior to their participation in the work, because ethics clearances typically specify users/researchers involved. **A complete safety check** should be performed in order to ensure that all project team members are up to speed on safety training and requirements in the project that is being pursued. **Safety guidelines** should be made available for everyone participating in the work. The guidelines should be placed in an easily accessible location, such as a binder in the research space or a digital file that can be accessed by all group members.

Recognition clarity

As the project commences, you should openly discuss **who will receive recognition** for their role and in what context. Projected authorship issues should also be openly addressed, and this discussion should be revisited and remain clear and open throughout the project, until it is completed and published.

Meetings and communications outline

A **meeting schedule/communications outline**, and guidelines for adhoc meetings and presentations, must be established at the commencement of the project. Will the team meet weekly, who will attend, and what will the nature of the meetings entail? Is the team expected to communicate intermittently, and will they work closely together each day or on certain days of the week? Some institutions have embraced "app" technology for communication between project participants located in different geographic locations (Appendix I). Will such apps be employed in the scope of this project? What are the expected lag times regarding communications, and who should be notified of different issues under various circumstances? Approaches for effective project communication will be expanded on throughout the remainder of this book.

Set up the budget and purchasing system

Once the project is greenlighted, you need to establish a solid budget plan, including implementing the systems by which you will access consumables and resources for the project, control purchasing, and manage petty cash. Most established groups will operate through an institutional budgeting system, but you may need to set up a separate in-house account for the new project. Spending of the assigned grant money should be followed closely to ensure the project is operating within budget, with monthly or quarterly reports that can be cross-checked by the project manager or group leader. You should clarify who will be responsible for managing the budget and clearing large purchases from the grant. Budget management is covered to a greater extent in chapter 4.

Setting Up the Team

<u>Defining the project team</u>

The project manager and group leader may need to develop the team at the start of the project by hiring from the new grant, or by bringing in an established staff member, collaborator, or student who is due to commence on their own scholarship. The best approaches to hiring are discussed in chapter 3.

<u>Set-up meeting</u>

Once the team is established, the project can officially kick off. There may or may not have been some preliminary work done already; in any case, everyone involved should schedule a meeting to make sure everyone is on track to move forward.

The project manager should prepare an agenda for this meeting to ensure that all of the critical points are covered. Similarly, the project manager should have a printed copy of the project plan for each contributor, outlining deliverables, deadlines, meeting schedules (if already established), expected project phases, the immediate next stage of work, and a summary of what has been achieved already. Communication is essential in a shared project, so providing access to shared project-specific resources is critical to minimize future hassle, because each participant should be able to access all of the information they need.

MATERIALS TO BRING TO SET-UP MEETING

- Agenda.
- Timeline (GANNT chart):
  - o deliverables,
  - o deadlines,
  - o meetings, and
  - o expected project phases.
- Outline for immediate next stage and participants.
- Summary of what has already been done.
- Information regarding project management shared resources (e.g. book, file, cloud-based app, or shared drive).
- Any important points regarding the project.
- Accessibility to resources, ordering requirements, permissions, and so on.
- Information regarding the expectations of each project participant.

EVERYONE SHOULD BE BROUGHT UP TO SPEED ON THE PROJECT AT THE SET-UP MEETING

Training Courses

Promoting skills and training is essential in many circumstances, but it also represents an **incentive for project staff**. You will observe a greater success in retaining skilled staff when you promote their professional development while they are working for you.

Several types of training courses may be completed at the commencement of a project. Some of these are essential, others recommended. They may be provided in house, or require budget to send project participants offsite to train. It is the responsibility of the group leader and project manager to define which specific skills will benefit the team.

Skills development

At the beginning of the project each participant, the project manager, and the group leader should determine any training courses that may be required to build the team skillset to be able to develop and complete the project. This may have been budgeted into the initial project, or may relate to training staff up on essential equipment, gaining required certifications, or undergoing basic skills development in new focus areas of investigation. In particular, establishing a new skill in-house (i.e. one that requires sending a team member on a paid training course) can end up saving expense over the course of the project and beyond.

CASE STUDY: Skills development

A new PhD student commences a doctorate studying the behavior of a marine species. The work will involve the collection of specimens with fieldwork day trips, which requires driving a car and a boat, as well as scuba diving. Upon commencement of the work, the student completes the required training as soon as arriving. By the time the project is set up and ready to go, the student is a qualified scuba diver, and has local driver's license and boat license. The project is able to continue smoothly for the student's candidature, thanks to the group leader's proactive forward planning which ensured that all of the required skills were established at the outset.

Induction training

If you are recruiting new staff to your institute to participate in a project, you need to ensure that they are fully operational as soon as possible in order to maximize their focus on the project once it is up and running. An essential part of this is **induction training** for your institution. This may relate to familiarization with the facilities that they need to work in, occupational health and safety induction, common resources, or student commencement programs.

Departments or institutions typically outline the required induction training for commencing in the department, but it remains the responsibility of the group leader to ensure that new project participants attend these sessions before they actively commence work. This is particularly the case if these participants have been recruited to join the project specifically. Establishing induction training in the first weeks of employment also presents a way to help the new person develop a professional network with those delivering and attending the training sessions.

CASE STUDY: Induction training

An early-career postdoctoral researcher commencing an international fellowship at a new institution was recruited based on their experience developing novel chemical compounds during their PhD. Despite their expertise in the field, they were required to undergo four days of local safety induction to the facilities on-site in their new host institution, and to receive a local operator's license to be able to operate the equipment required for the work they were set to pursue during their postdoctoral training period. Because this was established at the outset, the postdoc was able to progress immediately with their project, rather than waiting a long time for training to take place. The host group leader had organized all of the required training schedule to begin in the week the postdoc commenced work. This served a dual purpose of providing the postdoc with a networking opportunity with others pursuing work using similar techniques at the very beginning of their contract in the new group, and of promptly fulfilling the training requirements.

Management training

Although this book provides some useful guidelines for managing academic research groups and projects, it is a positive step to send your project leaders on management training courses to promote their skills.

Most PhD students and postdocs will not have received any training in management techniques, and their being made aware of approaches that may work to their advantage will promote greater project success in the long run, as well as improved professionalism, for all involved in the project. A number of generic project management skills courses are available online, or as several-day workshops, providing skills that can aid you in management as you develop as a scientist; further, they teach you about the potential management approaches that can be taken as you develop your own projects. Formal project management training will consolidate understanding of project management principles, in addition to adding value to the project leader's CV so they are more competitive for employment. Sadly, very few group leaders perceive the added value that comes from promoting professionalism through better management practices in their staff and students.

PROJECT MANAGEMENT TRAINING BOOSTS PRODUCTIVITY

Setting Up Facilities

Given the cutting edge nature of academic investigation, new projects will often involve the application of new technologies, many of which may involve the establishment of new facilities required to complete the outlined project. Although outsourcing can be useful in speeding up project outcomes, earlier chapters have discussed the benefits of developing facilities in-house that may provide a resource for the group to grow to become more productive and secure. Some projects will be dependent on the development of a new facility within the scope of the work, and the set-up of any new facility should be established during the commencement phase of the project.

CASE STUDY: Facility set up

A new PhD student has commenced the development of a novel transgenic mouse line in a university that lacks transgenic facilities. Whilst the central transgenic facility is under development and nearly ready to begin, the student needs to commence embryonic stem cell manipulation as soon as the transgenic targeting DNA is ready. The cell culture facility is established at the outset in collaboration with the project technician, and it goes through several months of optimization to avoid issues relating to infections, facility optimization, and protocol development. When the student is finally ready to use their construct, the facility is running optimally and the project proceeds with no delays. The facility goes on to develop numerous transgenic embryonic stem cells over the following years, establishing itself as a key resource in the institution and generating multiple shared authorships for the group leader who set it up.

This case study is demonstrative of how effective the establishment of a new facility at the commencement of a project can be, even if the facility is not required for several months. You must make time for troubleshooting in the set-up so as to minimize delays. For as many examples you may find of someone effectively developing resources in advance in preparation for later project stages, you will also find those who fail to prepare until the facility is actually needed. This results in overall delays in project progress, and it can be attributed completely to poor project planning and strategy.

Establish Work Spaces and Resources for Project Members
Each internal team member will need a work space for the project duration, and this should be set up at the commencement of the project. The type of space allocated will depend on the project focus; however, each team member should be given a desk space that they can use to process data, place orders, communicate, and write.

Maximizing the comfort of team members during the project, as well as facilitating their access to resources, will enhance productivity. Similarly, a clear outline for record keeping, including data storage and sharing for the project team, is an essential start-up task.

Establish Social Routines for the Project Group

In Section 1, in which we discussed the importance of social networking, the importance of social connectivity was emphasized in the building of a group and institutional network, but this is also a critical concern for project teams. People who go to work and spend the entire day working side-by-side with people they barely know are rarely happy.

Promoting social routines in the project team encourages greater happiness, and thus job satisfaction. Celebrating birthdays, key deliverable achievements, publications, or other social events can bring the team together and help them to function more effectively in the work space. If coupled with solid communication and professionalism in the workplace, these networks can evolve to span whole careers in the sector.

Start-up Checklist

A number of major start-up tasks have been introduced in this section. However, a key task for any group leader and project manager is to establish a simple start-up checklist that can be followed. This checklist should be specific to the project that you are developing and will differ slightly between each project, depending on numerous variables. For instance, if the work is ongoing from a previous project, many of the requirements will already be in place. An example of the key tasks that should be included in a start-up checklist are shown in table 11.1.

Table 11.1: Core project start-up checklist

Group leader (GL) tasks	
1: Initiate the project by confirming the project manager (PM).	
Project manager (PM) tasks	
2: Project manager conducts an update of the project plan and literature review.	
3: Project timeline and deliverables outlined on a GANNT chart.	
4: Together with GL, define the participants in the project.	
5: Make full safety assessment and guidelines available to participants.	
6: Ethics clearances for experimental work must be acquired, if not done already.	
7: Define who will be acknowledged and for what reason.	
8: Establish a meeting schedule and communications plan.	
9: Set up the budget and purchasing system.	

END OF CHAPTER 11 SUMMARY
Getting your project started

In this chapter we have covered the logistics related to launching your research project, particularly those focused on establishing your team. This section has outlined the need to set-up skillsets, prerequirements, and the workspaces required in order to commence the work. The key matters required in the set-up of your new project are covered here.

- Recruit a project manager and help them to establish their start-up tasks to set the project in motion.

- Permissions and clearances should be defined and applied for.

- Budget and purchasing systems should be set up at the beginning of the project.

- Oversee the project manager establishing the team and encourage the team to undergo any required training.

- Facilities and work spaces should be set up at the commencement of the project.

- The project manager should promote a social construct for the team, as well as define and follow a start-up checklist.

REFLECT

1: Have you operated effectively when starting up projects in your professional experience? Did the start-up run smoothly?

2: Consider how you have practically established project commencements in the past. Did you follow a start-up checklist? Was it effective?

3: Can you identify any extra tasks specific to your research that would routinely be included on a project start-up task list that have not already been included in this chapter?

EXERCISE 11.1: START-UP CHECKLIST

Use table 11.1 as a guide to build your own start-up checklist for the project you are currently on, or for an anticipated project that you have designed. Try to consider all of the key tasks that are standard in your discipline. You can download a table from the website (www.practicalacademic.com) to help you complete this task.

EXERCISE 11.2: PROJECT MANAGER GUIDANCE

As a group leader you are likely hosting several project managers, who are running several investigation foci. Consider the support that you provide in training your project managers to set up their work, and draft a guide that you can share with your project managers, outlining the key advice that you would like them to follow regarding setting up in your institution/group.

EXERCISE 11.3: START-UP TRAINING

What kind of training do you deliver to your staff and students at the beginning of a project, and do you consider the value-added? Make a list of all of your current staff and students related to projects, and list training that might improve their professional capacity. Are you able to finance this training, and will it improve on your budget or project deliverables?

Chapter 11 — Downloadable Materials

Download from www.practicalacademic.com

- Table 11.1 and exercise 11.1: "Start-up checklist" — Excel file containing a modifiable table 11.1.

CHAPTER 12

Managing the Project

Managing a project from initiation to completion requires a set of skills, which at least encompass good organization, people management, managing scope and expectations, technical expertise, and emotional intelligence. Good communication between all of the project team members and invested parties is essential for the cohesive progression of any project plan, and it brings in the valuable contributions of each team member. Each project-team member will present a skillset and knowledge background, which can bring innovation and diverse approaches to be able to progress the work and overcome potential obstacles with the minimum number of setbacks in project progression.

Types of Project

How you manage a project will depend on the type of project that you are working on. Several basic types and a brief description of each are noted here.

Solo project (SP)

Solo project management will pertain predominantly to scheduling both the project itself and your own calendar to pursue the work. This is typical for many postgraduate students and postdocs in smaller research groups, or in larger groups that fail to take on a cohesive approach.

The work will usually be overseen by the group leader and often an associate supervisor, and may include collaboration with other participants. The project is directed day-to-day by the individual that performs and directs the vast majority of the work.

Small group project (SGP)

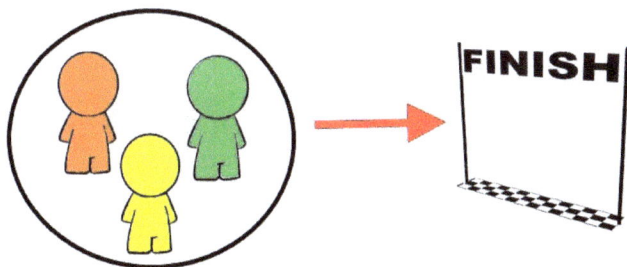

Small group projects may include any combination of active project contributors from within the group. SGPs usually include a postdoc, graduate student, and technician, or 2–3 people from within the same group. Members of the group may participate in other projects at times, or may only partly contribute to the main project.

The work is overseen by the group leader, who is usually the principal funding recipient. The project is usually operated in a close and collaborative style, and with the basic day-to-day guidance of the lead researcher in the team.

Team project (TP)

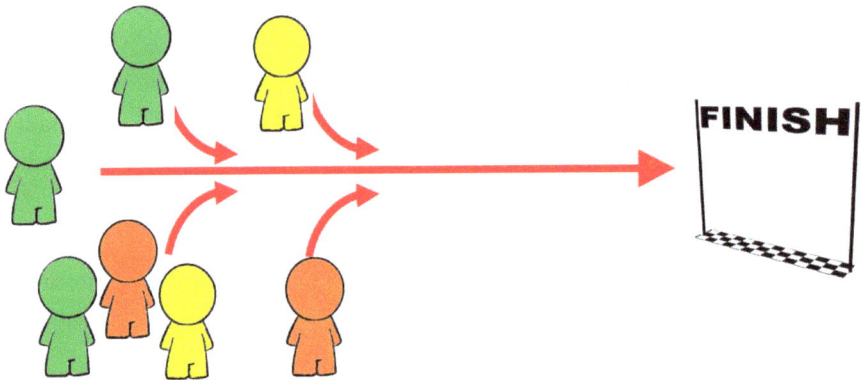

The team project involves a set team, but this may be larger than that of the small group project. It represents a major area of investigation within a large research group or between groups in the same department, and possibly encompasses several streams of investigation. Each project contributor may be actively pursuing other projects in tandem, and may only be required to contribute part of the entire body of work. This is often related to the contribution of a particular area of expertise. The project manager/leader is required to orchestrate the entire project involving scheduling and regular communication with invested parties.

The work may be under the overall guidance of one or several group leaders, who may be working together on a cross-collaborative or linkage project plan.

Large collaborative project / linkage projects (LCP/LP)

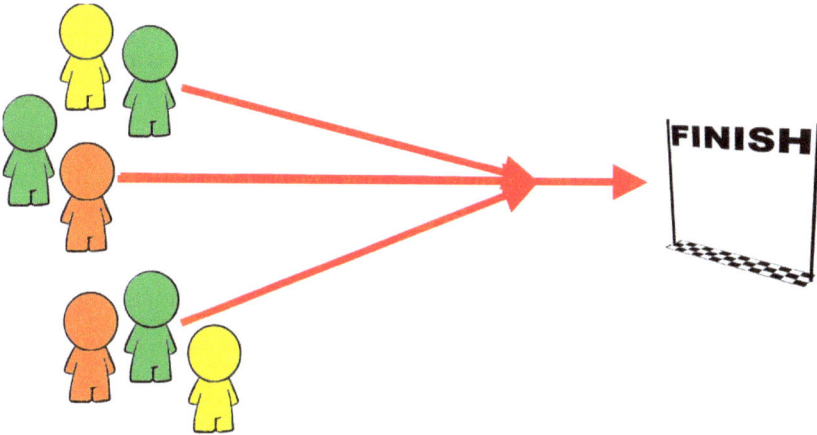

The large collaborative project will embrace intergroup collaborations, which often span different national or international groups, linked together through their project collaboration. It may include a regular flow of people and resources between the groups, and may encompass more than two groups in the process. These types of projects have become more common as funding bodies strive to bring together expert teams to address big research questions. Similarly, industry/academia collaborations are growing in number, as ways are sought to direct research to useful outcomes. In this type of project, there may be several project managers that work together from different home institutions to collaboratively deliver the project goals.

Multiple Project Prioritization

Favoritism

When multiple projects are taking place in your group, it is highly likely that you will have a deeper interest in some of the projects over other projects. As a leader, you must be aware of the issues that can arise from favoritism.

You may preferentially provide resources and support to your favored projects, driving their progress in advance of other projects also currently underway in your group. This can have a profound impact on your staff who are working on non-favored projects. Incentive declines when project progress is impeded because one's project is constantly being allocated to second place. It is critical that the overall group leader places equal interest and support on each of the projects being pursued by individual teams or researchers. Different projects will go through various stages of advancement, and **all of your teams and researchers deserve your support**.

Many group leaders fail to realize the impact that favoritism has on their group's work, and the desire to follow the most interesting line of investigation with full support will take precedence over other projects. However, maintaining systematic support and encouragement of other projects from your team is critical to your long-term success as a group leader. Following the guidelines offered here, including regular meetings and clear project planning and guidance, will ensure that you maximize each project's productivity under your overall leadership.

Project Team Communication

Clear and cohesive communication by all members of a project team, as well as by and with any external contributors to a project, is central to good project management (chapter 10). It aids project progression if all contributors remain on track. The best means of communication should be defined by the project team at the commencement of the project (chapter 11); however, some set approaches should include meetings and reports. The type of communication may also be dictated by the nature of the project, whether it involves extended periods of field work, whether a team member is remotely located, or whether the project is being conducted between dispersed locations. In these cases, communication may be predominantly via video conferencing, email, and telephone. More recently, digital apps have entered the project space as a faster way to stay in touch for project groups located in various locations. Common practices for project team communications are outlined in table 12.1.

Table 12.1: Approaches to project team communications

Type of project	Typical team communication	Communication strategies
Solo project	• Between PM and GL	• Direct communication • Email
Small group project	• Between group members • Between group and GL • Between individuals and PM/GL when postgraduate students are being mentored in the process	• Direct communication • Email • Video conferencing (remote members)
Team project	• Between group members • Between PM and group members • Between PM and GL • Between individuals and PM/GL when postgraduate students are being mentored in the process	• Direct communication • Email • Video conferencing (remote members) • Digital apps
Large collaborative project/ linkage project	• Between group members • Between PM and group members • Between PMs and GLs • Between PMs from different groups	• Direct communication • Email • Video conferencing (remote members) • Digital apps

Please see Appendix 1, where the digital resources for team communication are discussed.

GL = Group Leader; PM = Project Manager.

Meetings

Once the project is set up and the staff are hired, it is worthwhile establishing the meeting schedule for the group, so that you can plan out the practical schedule for the project to get underway.

Meetings are essential for **clear communication between members of the project team**. You should already have had a first team meeting, during which you will have discussed the overview of the whole project plan and the proposed contributions of each team member. As part of that, the regularity of meetings should be established, and these should take place throughout the duration of the project.

Weekly meeting

Weekly or biweekly meetings are recommended as a touch point for the entire research group, but individual project groups should normally have their own weekly touch point to go over how things are progressing — particularly if things are not working. This provides a way to keep all invested parties involved, and for those actively pursuing the project to obtain feedback about their progress and advice for future directions. For investigative projects, regular meetings provide a good opportunity to present project updates.

Checkpoint / stage meetings

In putting together your project schedule, it is worthwhile noting down where you might need to have special meetings to discuss the completion of a certain stage of the work, and to establish what the focus and approach is for the next stage of the work. These can be placed into the overall project plan at the beginning of the project, or decided upon as you progress.

Individual meetings

The regularity of individual meetings likely varies depending on your work type, but it is advisable that individuals working on the project meet with the project manager or group leader at least once each week to discuss the work being pursued, and to generally touch base. These meetings can take any guise, including Skype chats, phone meetings, or digital chats, if you are geographically isolated from each other. These may be included into the project group's weekly meeting, or may potentially be instead of it, depending upon how many people are working on the same project and upon its scope. Individual meetings aid the mentorship relationship, and ensure that project participants do not feel isolated and unsupported whilst they are focused on developing the work. From your point of view as the group leader, this is an opportunity to remain in touch with the research pushing forward in your lab, follow up on productivity, check the budget, and ensure the general wellbeing of your team.

Meeting structure

CASE STUDY: Preparing for meetings
A recent PhD graduate completed a postdoc at a highly-regarded research institution. After much trial and error, she adapted a system to achieve what she wanted from meetings with her group leader or with colleagues, which would minimize on the disruption to their day. She would simply ask for ten minutes, then she'd walk in, set the timer, and discuss everything she needed in ten minutes. This was a great approach to maximize free time for all involved, which is required for their own work and research, rather than being stuck in meetings. A key factor in the approach that this postdoc took was that she did her homework before organizing these brief meetings; she would sometimes spend two hours preparing all of the materials she wanted to go over in those ten minutes. Overall, this is a very good practice and a great organizational skill to develop in your students and staff.

Most of the meetings described here will take place over an hour or half an hour, depending on individual group preferences and schedules. But that is not absolutely necessary every time you need to meet with project colleagues. Quick touch-base approaches can maximize the required communication that needs to take place, but they restrict the time requirement over the long haul, thus freeing up more time for everyone to focus on their work. An example of this is shown in the case study. Preparing for a meeting minimizes the time required in the actual meeting; this is particularly useful during project phases, where time is limited.

PREPARE FOR MEETINGS TO MAXIMIZE TIME

Journal club

Junior researchers can often feel a huge amount of pressure when they are required to present a journal article at a journal club. It can be overwhelming to have to stand up and talk about a field that is relatively new, presenting to people that are at the top of their game and who are mentoring them. Nonetheless, finding and presenting relevant articles, be it for topic, technical reasons, or even simply because the article is outstanding, is a great opportunity for your project team to develop a collective mindset through academic discussion about topics relevant to you all.

Journal clubs can lead to new research ideas that can eventually yield high-level funding for research groups. Such funding can last for years. A journal club is a very useful tool to get your team working effectively.

Journal clubs promote the practice of presentation of new ideas in a public forum. That is an essential academic skill which must be developed, and it is best achieved by practicing in a supportive environment, within the scope of the group.

Journal clubs can promote the exploration of new ideas as part of a project team that are relevant to the project that is underway, to ensure that the group is in touch with the most recent advances made in the field.

Keep a log of journal clubs for all lab members to refer to when reflecting on the journal article material being discussed. This is also helpful for group members who were unable to attend the session. Updating the review log should be the responsibility of the person presenting. Presenting provides a good opportunity for team members to exercise their journal reviewing skills.

Table 12.2: Example of a journal club log

JOURNAL CLUB LOG
Date: 9th August 2016
Presented by: Joanna Briggs
Name of paper: Novel approaches to detecting volcanic rock deposits in the Auckland volcanic field.
Journal details: International Journal of Volcanology
Location of PDF file or print document: Shared drive/journalclubPDF/2016
Main context of work: New technical approach for detecting novel rock deposits.
Significant findings of paper: More cost effective and reliable technique demonstrated.
How could it impact what we are working on? Improve our analytical approach.
Overall impact on state of the art: Methodological improvement to expand on practical advances over time.
Notes:

Lab meetings

Lab meetings can be structured in a number of ways (see chapter 6), and they can be either formal or informal. Both have their merits. It is good to have regular group meetings to discuss general lab issues that arise. Oftentimes this is a platform for lab members to complain about other lab members' general practice, which should be something enforced through the chain of command in a different forum.

Collating the work to present on project progress every month or so in a formal way to the team or working group is a good way to make your team keep up-to-date on the current state of the art, and is a good opportunity to put their current data or research plans into a presentation format. This also drives your team members to develop the way that they present their work, which can be critically important for theses, articles to be published, and of course oral and poster presentations at scientific meetings.

Project meetings

Project meetings should be attended by those centrally involved in any given project, and may include participants from other groups or institutions. At the outset of the project, all participants should establish the frequency with which they should meet, and this should be included in the project overview at the commencement of the work (chapter 11). Given the dynamic nature of research, issues will arise, and participants may expand in number at times when the project may take on new avenues of interest, but the key participants from the commencement of the work should continue to guide the progression of the work. Project meetings should take place at least monthly, depending on the nature of the work, and should involve each invested participant presenting the current state of progress of their research. This gives everyone an opportunity to focus in on what will be addressed in the next month, prior to the next project meeting, and to ensure that everyone is on track. The management of expectations of results and outcomes, as well as any necessary troubleshooting between all invested parties, can take place in this way.

Meeting agenda

Whenever a meeting is called, it pays to develop a meeting agenda to be circulated in advance of the meeting to those attending it. A meeting agenda notifies all attendees of the topics that the meeting will focus on, providing a chance to prepare materials that may be relevant and useful to the team or group during the meeting. An agenda ensures that the meeting remains on track and that all critical matters are discussed in the time allocated. The agenda is usually developed by the person leading the meeting, which will typically be the project manager or group leader.

Table 12.3: Meeting agenda

Project meeting agenda
5th October 2016, Meeting room 1

Apologies:
Review of actionable items: from previous meeting.

1: General business

1.1   Facilities moving to new building for FACS

1.2   Revisions from recent paper submission.

1.3   Expansion of working space in department.

2: Reports

2.1   Current progress reports from each team member.

1.1   Collaborator reports.

1.2   Consultant/outsourcing reports.

3: Other business

4: Next meeting 6th November 2016

Record Keeping

Good record keeping is essential in all aspects of the work, including: planning; methodology description; data manipulation, analysis, and interpretation; and communication between project participants. The best practice for record keeping should be defined at the outset by the project manager and group leader (chapter 11), and must be followed for the duration of the project by all team members. Bad record keeping results in time loss and poor project progression at various stages, particularly relating to replication of the work itself or to the reporting and publishing of the work. More recently, good record keeping has become more essential for the development of intellectual property claims. Similarly, asserting your right to claim credit for work that you have performed is impossible if you have failed to keep reliable and clear records of the work that you have done.

GOOD RECORD KEEPING IS ESSENTIAL

Despite all of these facts, many academic leaders still fail to actively promote good record keeping amongst their staff and students, which represents a significant oversight in management practice. The approach taken will vary from group to group, depending on the nature of the research, but the group leader and project manager should actively define the expectations for record keeping at the outset of the project.

Some tips for good record keeping

- **Keep a bound day journal** where you can keep notes relating to the project, which are dated and signed off at the end of each day.

- **Keep a lab book** or experimental record book where you outline the work that you are performing. This can include notes relating to method in full, which is useful both when first following a new method, and for referring back to for repetitive experiments. This is typically a bound notebook, which is signed on each page. New technologies have adapted electronic record keeping alternatives that may be worth exploring.

- **Assign a secure folder on a shared drive** where each project participant can save their electronic data records, reports, references, and other project-related materials.

- **Request regular reports** overviewing the most recent advances in the work, which can be kept in a central project folder — both digitally and as physical printouts.

- **Photocopy/scan and keep a central store of all lab books.** Once an active experimental/lab book is full, it should be copied and kept in a safe location in case of the advent of patentable material or disagreement stemming from the work performed. The researcher should retain a copy of their lab book for their own records.

An excellent overview of best practices for record-keeping in academic research is provided by Schreier et al., (2006), which summarizes in great detail the types of considerations that must be taken when recording information in academic research.

Reports

Regular reporting by all team members should form a central aspect of communications, and it provides an opportunity for participants to keep up-to-date summaries for which the participant has taken the time to consider the project progress to date, and to make any relevant adjustments to the project plan.

Project report regularity

Report intervals should be defined in the set-up of the project; however, they will also be determined by the overall nature and duration of the project. Some general suggestions are outlined in the table below (Table 12.4).

In addition to these suggested time-restricted reports, detailed **end-of-stage reports** should be drafted by the team once a major component of the work is completed. These may come to constitute the basis for any patents or publications that stem from the work, and will serve to ensure that the work completed has no details that have been overlooked, which would require the team to repeat experiments later.

Table 12.4: Time-restricted reporting on projects

Project length	Project type	Major reports	Brief updates
3 year research project	Complex, several investigation foci	Each month Each quarter Annual	Each meeting
2 year research project	Fast, typically 1–3 research foci	Each month Each quarter Annual	Each meeting
9–12 month research project	Fast, typically 1–2 research foci	Each month	Each week Each meeting
5–7-week summer research project	Short, low risk, mentored	End of project	End of week

Report structure

You may have an established report structure for your particular field, which will represent the best practice for reporting. However, in general you might build your project reports on the classic research style, but written with more brevity for simplicity. This type of report structure allows the project manager to best follow the progress of the project as it proceeds.

Simple introduction — to outline the relevance of the work being reported on relative to the whole project plan.

Method — the method being employed for the currently-reported work, in a format similar to that which would be reported in a journal.

Results — tables and figures outlining the data being reported for the current reporting period. These are coupled with a brief results summary, as would be noted in a journal article.

Discussion — relating of the results back to the simple introduction.

Challenges and issues — note of any challenges, setbacks, or issues that have been confronted in this part of the project.

Future directions — explanation of what will take place in the next stage of the work.

Appendices — attach any support data that is required for readers to understand the overall project, so that these results can be understood.

Activity log

The project manager should consider keeping an activity log, which in essence may be a day journal, where notes can be recorded regarding project-related issues and ideas. A clear, updated summary of the work underway, key operation information, and a projection of the direction in which the work is moving may be recorded here. Any day-to-day troubleshooting concerns, key results, or unexpected findings should be noted in the activity log.

Keeping an activity log ensures that the project manager has an informal reference carrying key information to keep track of the project and its dynamics during the action stage of any project. The activity log is particularly important when unexpected verbal updates regarding the current project are required.

Lessons learned log (LLL)

A "lessons learned log" should be kept by the project manager to keep track of the components of the project that proceed well, or which fail. The details of how to develop and manage a LLL are outlined in chapter 14. This may extend on from an activity log, but it has a greater focus on improving your practice by identifying what works well and what should be improved upon as the project progresses.

END OF CHAPTER 12 SUMMARY
Managing the Project

In this chapter we focused on issues relating to managing the active project. The main topics covered project types and their prioritization, communicating, meetings, and record keeping. The main take-home messages include:

- The type of project that is being pursued influences how the project is managed. This particularly impacts the team structure.

- The projects underway in your group should all receive support from the group leader, regardless of stage or current advances.

- The project team should maintain an effective communication regime and participate in regular meetings.

- Clear records should be kept by all project participants, regular reports made, and the appropriate journals/logs kept.

REFLECT

1: Are you organized in the way that you manage your active projects? Do you establish regular meetings and touch points to ensure that everything remains on track?

2: Do you exhibit favoritism when supporting the advancement of different projects in your group? Does this impact the members of your team who are not focused on your preferred project?

3: Do you ensure that your project managers encourage regular reporting and keep an activity log, to stay on top of project progress and issues confronted?

EXERCISE 12.1: COMMUNICATION
Make a list of the projects currently underway in your group and define them according to the categories described in this chapter. Note the types of communications that you promote between your project teams, and also those you use yourself. How do your teams stay connected?

Project details	Communication type	Between whom?

EXERCISE 12.2: MEETINGS
Establish a projected meeting schedule for each project team that you currently have working in your group. If you operate an individual research project approach, design a meeting schedule for the whole team to network over each other's projects. Consider the regularity and type of meetings that you will schedule.

Project	Meeting type	Regularity

EXERCISE 12.3: REPORTS

Design a report structure, including regularity of reporting and key stages at which reports are required for each project underway in your group. Include to whom the reports are delivered and the mode of delivery (oral/written).

Project	Report type/details

Chapter 12 — Downloadable Materials
Download from www.practicalacademic.com

- Table 12.2: "Journal club log" — Word file containing a modifiable table 12.2.
- Exercise 12.1: "Communication" — Excel file containing a blank template and an example of the completed exercise.
- Exercise 12.2: "Meetings" — Excel file containing a blank template and an example of the completed exercise.
- Exercise 12.3: "Reports" — Excel file containing a blank template and an example of the completed exercise.

CHAPTER 13

Troubleshooting and Change

Research projects are prone to change, especially if they span more than a year in duration. Any number of reasons could trigger the need to change. Several of these reasons are discussed here, followed by approaches that you can take to formally adapt your work.

Being out-published

Given the highly competitive and cutting edge nature of much research, combined with the peer-review process and international scope, you may oftentimes find yourself years into a large project, where you can see the finish line in sight and when you have a wealth of data, when another group beats you to the punch and publishes the work. It would be foolish to carry on with your project as planned, given the exact data you are aiming for has now already been published by another group. In this situation, you need to look at their work with a fine-tooth comb, gauge where they are taking it next, and compare it with what you've already done. All is not necessarily lost, and you can actually use this opportunity to take your work further.

Your project team will no doubt be devastated by this turn of events, but it is not the end of the world. The first thing to do when this type of situation arises is take some time to think about how this impacts your work and how you could move forward from this point. Encourage the team to brainstorm for options regarding how the project could be adapted to achieve useful or novel outcomes from the already-completed work. Did the competitors address the same questions in exactly the same way? Can the completed work complement what they've done with a slightly different focus?

ASSESSING AND ADAPTING TO BEING OUTPUBLISHED
1) Review the published work,
2) Meet together to brainstorm possible approaches,
3) Define possible approaches to progress the work,
4) Redesign project plan to incorporate change
5) Adapt all relevant processes to accommodate change.

ADAPTATION TO CHANGE IS KEY TO MOVING FORWARD SUCCESSFULLY

New technology
Research can be notoriously dynamic, especially in experimental research, and new technologies are becoming available every minute of every day. Similarly, technologies are becoming more affordable. New technologies can dramatically speed up your work, too. The process of adapting to new technologies is similar to adapting to any other change. There are several steps that you can follow as a researcher which will help you and your team to adapt and move forward with the incorporation of the change.

ASSESSING AND ADAPTING TO NEW TECHNOLOGY
1) Acknowledge the availability of the new technology.
2) Complete a cost evaluation analysis of the new technology, outlining the benefits versus the change in cost.
3) Acquire or access the new technology on a trial basis to evaluate it.
4) Adapt all relevant processes to the new technology.

CASE STUDY: New technology

During a one-year Honors research project, the host university welcomed a new in-house automated DNA sequencing service. The student commenced the project using radiolabeled nucleotide-based sequencing gels (which is time consuming, expensive, and dangerous), but had adapted to overnight automated DNA sequencing by the end of the project. This new, faster, and safer method was run in tandem with radiolabeled sequencing for much of the project to confirm its effectiveness and reliability.

New collaborations

Through networking (discussed later in this chapter), you might come across new potential links with groups that already have something you had planned to create yourself. This could be a reagent, an algorithm, a technology, a piece of DNA, or a particular mouse strain. All manner of useful materials can be sourced through effective collaboration, and these things can boost the productivity of your group's work and, in particular, can boost your project's progression. Something you had outlined to develop during your project might be accessed through collaboration, essentially cutting time out of your project timeline and freeing up more time for project advancement.

Although new collaborators may contribute to the publications that your group creates downstream, the core project group still needs to incorporate this change and update project plans to accommodate it.

ASSESSING AND ADAPTING TO A NEW
COLLABORATION

1) Acknowledge the availability of the new collaboration.
2) Complete an analysis of how the new collaboration will benefit the current project.
3) Discuss the potential of a new collaboration with your project team.

4) Identify any potential sources of conflict from this collaboration that may impact your project team.
5) Discuss the new collaboration with the collaborator.
6) Establish groundrules with the new collaborator.
7) Commence the new collaboration.
8) Adapt project plan and all relevant processes to new collaboration.

A project or group leader might rush into a new collaboration out of excitement for the new potential project advancement, but caution should be taken to establish the ground rules. You need to identify the impact that this will have on current project contributors, particularly in relation to recognition once the work comes to publish (see the case study entitled "collaboration," chapter 10). Similarly, you need to identify who will have the rights to continue this project once the work is published. If an author's name is listed on a paper, they may consider it their right to continue on that line of investigation moving forward.

Key project contributors leave
Whether due to health, work, personal, or other reasons, it is likely that project contributors will leave during projects throughout your academic career. No matter their role, this impacts the project and requires the attention of the project team to address how to best navigate the change.

Group leader moves mid-project
Academic careers can be dynamic, and if the group leader excels in their field, they can often receive offers that they cannot refuse from outside institutions, often internationally. For established group members, this may significantly disrupt research progress, and the team is usually expected to assist in the move, organizing equipment to send and setting up the new research space in the new institution This can also impact enrollment for research students, whilst students and postdocs will need to address issues around their scholarships and

fellowships. A group leader moving mid-project can be a verifiable nightmare.

For students and staff mid-project:

- Find a host group in your current institution that might support completion of the current project under co-supervision with the migrating current group leader.
  - This option is important for those who are settled, for personal reasons, in the location of the original institution.
  - This option is preferred when the new institution provides no advantage to project completion, or when the work is almost complete.
- Re-enroll at the new institution, gaining credit for the time already completed, potentially with new funding from the new host institution. Continue with the move as a team member.
  - This will provide experience in moving projects between spaces and setting up research space, which may come in handy when commencing your own independent academic career.
  - This works best when the new institution is more renowned than that which you are moving from.
  - This will impact project progress, but may invigorate the outcomes with access to a wider research network and more advanced resources to progress the work faster at the new institution.
- Complete the project write-up in the current institution, and then continue on with the next follow-up project in the new institution.
  - This works well when the project is almost complete and perhaps only a few experiments are required to complete it.
  - Papers and theses can be completed in the original institution in assigned office space, and the student/postdoc may then continue on to join the group for a follow-up project at the new institution.

The best-case scenario for group leader migration is where the group leader manages to negotiate an overlap period, in which the original group space is kept for a period of time under the direction of a senior postdoctoral fellow. This would allow the team members in the original group to complete their projects and look for their next role with less pressure. In tandem, this would also give the postdoc a brilliant opportunity to develop their leadership skills, which may result in prospective career advancement.

Dealing with Project Change

Depending on the type of change that you are confronting, you may consider putting projects on hold, if possible. After discussing with individuals regarding the potential impacts this change may have on them, you should announce the change you are confronting to the rest of the project group, and request a special meeting to discuss the project and where it might go. In doing so, you should request the project members take some time to consider their future directions also, so that you might all consider the situation, each bringing to the table their own idea about how the change should be addressed. Once you've decided on a course of action together, you need to determine whether the project change is significant enough to notify your funding body of the shift in project direction.

The core of change management, and the associated project change methodologies, is similar to that which you may encounter in a business setting, but the flexibility of the academic environment means that you will have different dynamics come into play with regard to the people involved. A summary of common types of project change with suggested adaptation approaches is shown in table 13.1.

Table 13.1: Adapting to project change

Reason	Issue	Possible approaches
Out-published	Your work becomes redundant because someone else publishes it.	Determine aspects of your work that extend on from that published to maximize your potential output. Re-structure your project plan.
New technology	Project may benefit from adopting newer technology than what is already used.	Reorganize project to accommodate the new technology available, and restructure the project plan accordingly.
New collaboration	A new collaborator that could advance the project's progress significantly is found.	Establish the ground rules for collaboration before beginning the collaborative pursuit.
Group leader relocation	The group leader has taken on a new position in a different location. This involves moving the research space, disrupting work progress.	Depending on the stage of project, the researcher may stay back to complete the work, or may adapt their work to take place in the new host institute, with renegotiation of funding and contracts.

Checklist for Project Change

An example of a basic checklist for dealing with project change is shown in table 13.2. This checklist outlines some key questions that you can pose in order to help navigate the change that you are confronted with. This example deals with a potential new collaboration; however, this approach may be taken for any type of change confronting your project.

Table 13.2: Model of checklist for project change

What is the reason for the requirement of project change?	New collaboration can promote advancement in project progress.
How does this impact the project that is currently underway?	Outsourcing key stage to collaborator, inclusion of more project participants.
Who is involved in this project and what are their roles?	Postdoc (JG), PhD student (TA), technician (NM).
How will this change impact the project participants? How can this be addressed? Pros/cons?	Potential impact on authorship. Student may join collaborator to complete the work offsite, expanding their network, which may reduce the authorship debate.
Potential solutions to work around problems posed by the change.	Open communication regarding reframed project and progression from here.
Changes in costs and time to impose each potential new solution, and how it impacts the overall project time, scope, and budget.	Reduced costs and enhanced time. Opportunity to expand on side project whilst other work being completed offsite.
How will you manage the materials that you have already developed in the project thus far?	Ongoing in the same manner, but expanding on digital networking regarding data.

It is worthwhile completing a checklist or report outlining the changes to your project. This ensures that everyone is on the same page. If the project is being performed as part of a thesis project, this provides material that can be included in the thesis. If you are reporting on the progress to a funding body, this information will be available to report to them in the next progress report period. The key is to think logically and research the ongoing direction that the project will take.

Commencing new project avenues / changed project
Given that the project can expand in new directions, the processes required for the setup of a new project may need to be started anew in order to continue to progress the work in these new directions. These establishment processes are outlined in chapter 11; however, in brief, they include:

- involving the project team;
- stating the project timeline in GANNT chart form;
- obtaining required permissions and clearances;
- establishing a budget;
- establishing a communications strategy;
- establishing a meeting schedule;
- defining and setting up required training;
- setting up required new facilities; and
- establishing required new work spaces.

Enhancing Troubleshooting

Given that the nature of academic research is to pursue cutting-edge innovation to advance the current knowledge of a given subject, setbacks and issues are common. Thus, a central component of managing academic research projects is being able to **confront setbacks to keep the project moving forward**. This might not require such significant adaptation as outlined above in the discussion on project change, but instead might be as simple as adapting to changes in reagents available, the breakdown of essential equipment, or the contamination of a key sample that is currently being studied.

Experienced project leaders are practiced at troubleshooting, and this is a skill that must be embraced and taught in the mentoring process. Several common approaches to improving troubleshooting are noted here.

1) **Establish a local research network**, and familiarize yourself with the accessible resources in your current location. You can petition your network for resources that you might borrow whilst waiting for your order to arrive, or you might draw on the collective expertise of others to provide suggestions of how to deal with the issues you are confronting.

2) **Be aware of the resources available to you**. The prior knowledge of where potential substitutes are available will limit the delays that you encounter when something fails. This may include a piece of equipment, reagent, or facility.

3) **Consider several approaches that might be taken** to address your current research task when you are designing the work. If a stumbling block is confronted, you may already have a backup option ready to go.

4) **Engage the project team and research group** to help in the troubleshooting process. Multiple-brain power is stronger than singular. Employ your team and initiate their troubleshooting skills to complement your own.

5) **Have another team member approach the problem.** Sometimes your protocols will be sensitive to individual approach, and researchers may be inadvertently doing something wrong in their technique. Handing a protocol over to an experienced colleague to give it a try can sometimes resolve the issues. Observe the colleague in operation and see whether the error is apparent. If both cannot get the work going, user error can be omitted.

6) **Observe others using the technique.** Similar to point #5, by observing others using the same technique you might resolve where you are going wrong in your technique.

7) **Outsource the work.** If all else fails, identify whether you can get this part of the work completed by outsourcing through service providers or collaborators.

8) **Replan the project.** If you cannot resolve the progress of a key project stage through any of the above troubleshooting approaches, you might consider dismissing this project stage and making a change of plan altogether.

Managing Project Time

A focus on time management is made clear throughout this book as being a very central component of effective research project management. Being able to manage resources and people to deliver the required results within the timeframe outlined is particularly important with regard to project delivery. Although most academic funding backers understand the volatile nature of discovery research, reaching landmarks of an outlined project within good time is indicative of effective project planning, management, and delivery. Different techniques are discussed throughout this book regarding time management; nonetheless, in the scope of managing the active project, this relates mostly to **effectively troubleshooting, adapting to change, and ensuring good team communication** to ensure everyone is on task. The best practice to address these factors is discussed within the scope of this section.

At the commencement of a project, the lead researcher/project manager should establish the schedule of planned project outcomes, and the timeframe in which different phases of the project should be pursued (chapter 11). We've already discussed here how these are prone to change, thus the original plan is likely to go through multiple adaptations during its path from initiation to delivery. Similarly, the daily schedule is likely to be dynamic for most project participants, anchored by routine meetings and any other extraneous commitments.

The only way to manage the daily, weekly, and monthly schedule is to plan out specific times each day to accommodate project tasks and deadlines. Set some time aside each week to review your current planned schedule.

<u>In your regular schedule revisions you need to:</u>
1) Consider events that may impact your project schedule (personal reasons, conferences, teaching, change, and so on).
2) Adapt and revise your calendar to accommodate these events.
3) Update the project GANNT chart.
4) Communicate relevant significant changes in the project schedule to other participants.

By actively setting aside time to reflect on your schedule, you can keep on top of the project plan and be capable of communicating the current timeframe and expected deliverable dates.

GOOD TIME MANAGEMENT PROMOTES IMPROVED PRODUCTIVITY

Practically speaking, you need to be aware of which scheduling approach works best for you and your team: diary-based, electronic calendars, spreadsheets, or group scheduling tools. Everyone has a different style. Write a summary of the principal recurring responsibilities that you have each week and how they fit into your schedule. It helps to keep a paper day diary that you can update as commitments become apparent, and these can later be transferred to electronic schedules. Actively developing effective time management skills at the outset that can persist throughout the project will only aid effective project completion.

Keeping Up-to-Date with the State of the Art / Literature

Time should be taken to keep up-to-date with the current state of the art throughout the lifetime of the project. New studies can come from left field and change the entire project focus from one day to the next. Similarly, keeping up-to-date with the literature helps the project participants to stay well informed during the project. Communicating any useful new studies to the team can maximize cohesiveness and provide a focal point for meeting discussions. Try scheduling time each week to check the changes to the state of the art. Conference proceedings can always help to stay on top of recent advances relative to the current investigation topic.

Project Networking

Networking in itself is an important skill that is touched upon repeatedly in this book, but here it is spoken of in the sense of project progression. Networking within the scope of the project can impact the project you are pursuing in immeasurable ways. It is impossible to predict how networking can specifically impact different aspects of your work, but without networking no related opportunities will present themselves; thus it is advisable to develop solid professional networks wherever possible. Networking is an essential part of any good researcher's life.

Project-focused networking is not restricted to conferences. Chapter 5 outlines a range of options for establishing professional networks — this includes conferences, departmental activities, institutional networks, and online networks.

At conferences

When delivering your most recent project outcomes at conferences and meetings, you are encouraged to network with people in your field or with those that use your techniques. It is quite likely that your group has established networks, but conferences are an opportunity for you to build your own networks further. It is a good idea to take some time before attending conferences to look at the abstracts from the other groups to gauge the work they are doing and the techniques they are employing. Similarly, have a look at their websites, where available, and gauge if there is anything at all of interest that you can discuss with different people at the conference. Keep a notebook with you; note down the groups and the people, and the work you want to talk to them about. When you get to the conference you will be able to immediately start a professional conversation when you come into contact with the people whose work you have read about, which is particularly important when it comes to attending oral presentations and poster sessions.

Networking to source resources
Do not be afraid to contact other groups when you know that they have resources you might need for your project. Email or phone calls, depending on the time pressure, are both acceptable ways to contact other groups relating to sourcing materials.

When trying to source resources, have a clear outline of the following before reaching out to a potential resource provider:

- What exactly you are seeking, what do you need it for, and how are you going to use it.
- How will you acknowledge the group that is sharing the resource, and are you willing to include their names on your papers?
- Would you prefer to send someone from your group to use the resources the potential provider has, or simply have the resources sent to you?

If this rolls over into a collaboration, it may evolve into funding applications and years of working together. Never underestimate what might evolve from reaching out for access to a resource.

Digital Resources for Networking and Project Management

Managing your project can be made easier by keeping a communal area that is accessible to all participants and in which contributors can store data. In this respect, cloud computing has become more commonplace in the academic research sector. Issues relating to this pertain predominantly to intellectual property security. When jointly authoring manuscripts or sharing data, files that are more recent can easily be shared through a Google Drive or Dropbox folder, or a communal folder on another shared drive system. What works best for your project team will depend on your individual setting. Being able to share resources will maximize the team's ability to coordinate their work, particularly regarding keeping up-to-date with project advances and changes that take place at any given time. A number of relevant digital resources for effective project management are noted throughout this book. These are covered in more detail in chapter 14 and appendix 1.

END OF CHAPTER 13 SUMMARY
Troubleshooting and Change

In this chapter we discussed how best to manage change in your projects in response to various common issues that arise. Best practice on improving troubleshooting, adapting project schedules, and networking to enhance adaptive capacity are also covered. These topics were delivered as follows:

- Various scenarios commonly trigger change in academic research, and some best-practice approaches are discussed in this chapter.

- When you encounter the need for project change, a checklist for project change can prove a useful tool.

- When the project leads to novel research directions, it needs to be re-established following the guidelines in chapter 11.

- Various approaches are described to help enhance your troubleshooting capacity.

- Keeping up to date with the latest advances and with networking can help to stay at the cutting edge in your project approach.

- The project schedule must be regularly revised in order to retain realistic project timelines, and digital resources may be employed to manage more effectively.

REFLECT

1: Have any of your projects suffered a major setback requiring a substantial change to the work being pursued? How was it addressed?

2: Consider how you might best improve your capacity to troubleshoot stumbling blocks in your work. Do you follow the guidelines listed in this chapter? Do you have other approaches?

3: Do you and your team keep up-to-date with the current literature in your field during the lifetime of a project? How do you schedule this into your busy week/month?

EXERCISE 13.1: CHECKLIST FOR CHANGE

Employ table 13.2 to create a checklist for project change on a project that is underway or which has previously happened in your research. This blank template can be downloaded from the website (www.practicalacademic.com).

What is the reason for the requirement of project change?	
How does this impact the project that is currently underway?	
Who is involved in this project, and what are their roles?	
How will this change impact the project participants? How can this be addressed? Pros/cons?	

Potential solutions to work around problems posed by the change.	
Changes in costs and time to impose each potential new solution, and how it impacts the overall project time, scope, and budget.	
How will you manage the materials that you have already developed in the project thus far?	

EXERCISE 13.2: TROUBLESHOOTING AND CHANGE

Make a summary of some major changes that have impacted your research projects during your career. Now summarize the approaches taken which addressed that change, and note how effective they were. Contemplate what could have aided a better response to these issues.

Change	Approach	Outcome/effectiveness

EXERCISE 13.3: PROJECT NETWORKING

List the current networks that you are a part of which might impact the efficacy of your team to respond to change and to troubleshoot issues. Link these networks to particular types of assistance that you can gain from participating in a networking event.

Network	Specific assistance

Chapter 13 — Downloadable Materials
Download from www.practicalacademic.com

- Table 13.2 and exercise 13.1: "Checklist for project change" — Excel file containing a blank template and an original table 13.2 as a reference.
- Exercise 13.2: "Troubleshooting and change" — Excel file containing a blank template and a completed example.
- Exercise 13.3: "Project networking" — Excel file containing a blank template and a completed example.

REFERENCES – Chapter 13

Schreier AA, Wilson K, Resnik D. Academic Research Record-Keeping: Best Practices for Individuals, Group Leaders, and Institutions. *Academic medicine: journal of the Association of American Medical Colleges.* 2006;81(1):42-47.

CHAPTER 14

Communication and Improvement

A number of factors need to be taken into consideration when it comes time to communicate and reflect on the work that has been pursued during the project. These matters are discussed here.

Intellectual Property (IP)

The ownership of intellectual property is a key concern in many academic research projects. Financial reimbursement from well-developed intellectual property can make your career, depending on how long your work is adaptable for financial gain. Patents cost a lot to file and prove, so the group leader has a vested interest; however, the work could not be pursued without the intellectual contribution and participation of the entire team working on that project. This presents a significant issue when working on a short-term contract that delivers patentable technology. Established teams tend to protect their intellectual property and ownership of the technology, often at the expense of collaborative growth. **Defining ownership at the outset** — at the beginning of your work — is the only way to circumvent or minimize such issues that you may come across later in the project. Thus, as a new project commences, it is critical to make an agreement regarding IP. It is entirely possible for the group leader to "buy out rights" if a group member decides to leave the group, which is commonplace in various research sectors, but oftentimes lack of communication makes the whole situation challenging.

CASE STUDY: Intellectual property ownership

An industry-focused research group had developed lead small-molecule therapeutics, but their biological efficacy had not yet been proven. A postdoc was recruited to aid the work on short-term contract. The group leader failed to discuss IP rights over the work, and chose not to integrate the postdoc into the established team. Nonetheless, the postdoc envisioned multiple possible research directions stemming from the work and was encouraged to develop a new line of investigation. During this time, the group leader would come to the post-doc's desk each morning with a coffee in hand and say, "you are going to steal all of my ideas and go and start your own lab." After the new project was running well and poised to produce great data, the group leader insisted that the work be passed to a permanent staff member, and the postdoc was encouraged to come up with another new project. For all intents and purposes, the postdoc had become a "new ideas generator" for this research group. After six months, the second project had been taken from the postdoc, who began to appreciate the gravity of the situation. The postdoc left the group some months later, moving on to another, more successful role, but leaving a gap of almost a year on their resume, from their time pursuing patentable technology, with nothing to show for it.

This case emphasizes how focusing on intellectual property generation can impact research groups. All innovation must be retained as a secret in order to be able to patent the work. It majorly limits your capacity to network with outside researchers to push the work forward. Nonetheless, patentable technology is necessary to push advancements forward. It is absolutely essential to communicate openly regarding the potential impact that seeking patents will have on each project participant, particularly those on short contracts or those at an early stage in their professional academic career. This clarity allows the project participants to gauge whether they are happy to continue on the project, when doing so may mean they are unable to boost their profile with journal publications for some time. Alternatively, they may prefer to move on to another role where they can focus on publishable work. It may also be viable to offer your team working on patentable research some alternative side projects that can yield publishable research to help keep their track record competitive.

COMMUNICATING WITH YOUR TEAM ABOUT PROJECT GOALS IS ESSENTIAL WHEN WORKING ON PATENTABLE TECHNOLOGY

Communicating with the Funding Body

Most funding bodies request intermittent reports to keep updated on the progress of any research project, most critically to gauge whether the project is progressing according to plan. Progress reports will vary in content from funding body to funding body, and project to project, dependent on the established practice for that funding provider. It is critical that the project team be able to **deliver clear updates regarding the current project achievements**, and be able to argue the reason for any setbacks that may have delayed reaching set deliverables on schedule. Special reports may sometimes be submitted when projects suffer significant setbacks or major changes in direction (see chapter 13), which may require reassessment by the funding body. Funding bodies usually require end-of-project reports, which will be covered further in chapter 16. The regular meetings and solid record keeping advised in chapter 12 make the collation of reports to funding bodies easier to address efficiently.

Communicating Your Outcomes

Once your project group achieves key deliverables, it is time to communicate your work. In academia, this is predominantly done by publishing your work in preferably high impact, competitive, and peer-reviewed journal articles. For students, a thesis is likely the top priority. In any case, each project participant should identify what their primary goals are to achieve from the work. Examples of these goals are noted in table 14.1.

It is interesting to note that most of the participants in the project benefit in the same way, except for the technician who likely will not be presenting the material at conferences nor getting their name on review articles. This emphasizes the benefit from working in cohesive teams to achieve common goals that usually align well. Success of the team benefits all participants. It is also good to be aware of the benefit that each participant is gaining from the project, to be aware that you are never truly working only for yourself in experimental academic research, and that each of the invested participants has a role in the game.

> ## EACH PARTICIPANT MAY BENEFIT DIFFERENTLY FROM PROJECT OUTCOMES

Table 14.1: Typical project goals for participants

Group leader	- Publication in high impact journal. - Intellectual property for market. - Resource creation for market or future projects. - Successful mentoring of postgraduate student or postdoctoral fellow. - Review article derived from project overview. - Conference presentations and invited speaker.

Postdoctoral fellow	- Publication in high impact journal. - Intellectual property for market. - Resource creation for market or future projects. - Successful mentoring of postgraduate student. - Review article derived from project overview. - Conference presentations and invited speaker.
PhD student	- Publication in high impact journal. - Intellectual property for market. - Resource creation for market or future projects. - Review article derived from project overview. - Conference presentations.
Technician	- Publication in high impact journal (some cases). - Intellectual property for market. - Resource creation for market or future projects. - Successful mentoring of postgraduate student or postdoctoral fellow.
Collaborator	- Publication in high impact journal. - Intellectual property for market. - Resource creation for market or future projects. - Conference presentations.
The host institution	- Intellectual property rights. - Journal publication output. - Future funding potential. - Increased appeal of the institution. - Publicity potential from project successes. - Increased institutional profile. - Student graduations.

The funding body	- Successful research productivity in funding body's name. - Improved state of the art in focal research area (e.g.: breast cancer). - Publicity potential for funding body.

Project Presentation at Conferences and Meetings

Here we revisit conferences and meetings from the point of view of **delivering a successful project outcome**. Anyone that has their name listed on a conference poster can regard that as a communication of their work that they can list on their résumé. Thus, it is important that anyone contributing to the work being presented at a conference, workshop, or meeting has their name in the list of authors. This is discussed further in chapter 15. If the presentation is representative of work that is yet to be published, it can be considered to be the interim presentation of the work that will eventually be published; as such, all the names that would go on the paper so far should be included in the list of authors, with the presenting author noted. Being a first author or corresponding author is not required to present at a conference or meeting.

The decision of who to send to conferences or meetings from a research group may relate to funding. If you are attending as the group leader, sponsored by your department, then do encourage your staff or students to attend. Most often they will be very keen to get the opportunity to network or see what other groups are doing. Similarly, two heads are better than one, and you can gain a lot of extra insight from your group members.

Look at the possibility of planning attendance at talks together, maximizing your exposure to new information through coordinating which talks you will attend. Consider training your group members in what you expect of them when they attend talks, what types of notes to take, and what kind of information you want from each abstract presentation (see more on preparing for conferences in chapter 5). If there are multiple relevant conferences each year, it would be wise to stagger attendance between the research team members.

Publications

Similar to conferences, the publications that stem from the project are of importance to most people that are involved in the work, with the exception of paid contractors. Who will be recognized when it comes time to publish the work derived from the project is discussed more extensively in chapter 15.

Project-derived Technology Development

Technology is often developed by advancements in projects undertaken in the academic setting, and how this technology is adapted can vary. Patents are discussed earlier in this chapter; however, the technology may not only present a marketable commodity. Technology may also render future pursuits in the field to be more achievable due to technical improvement, novelty of approach, or innovation in detection methods, for instance. For this reason, the technology developed from a project can set up future project success, and such technology can be employed by groups to aid the advancement of follow-on work.

The key concerns when new technologies are developed in this setting relate to the rights to use or adapt the technology for ongoing work. Given the nature of the academic sector, junior researchers often move between institutes and projects easily, and they may or may not be permitted to take the novel technology with them — particularly if they are going to work for a competitor. Understandably, group leaders often prefer that the technology remains in their hands and not in the hands of their peer competitor in the field. Who owns the technology is a matter to negotiate with the research team, and a matter to establish as an open discussion from the outset.

Failed Communication and Discontent

As discussed in Section 1, in which we focused on groups, teams sometimes also experience stages of discontent where one or more project participant is dissatisfied with another in the team, or perhaps the overall direction of the work does not please all team members. These types of issues often come about from a **failure to establish a clear and open communication practice** when the work commences. Numerous issues can arise.

Project participant discontent has been known to relate to dissatisfaction with the project direction, the overall management (or lack thereof), the intellectual input, a lack of goals or a sense of floating, a lack of transparency with all team members, dictator-like control over the work, unethical management, or being undercut on your contribution.

CASE STUDY: Team discontent

A junior researcher joins an established team to pursue a one-year project to develop technical approaches to diagnose a complex genetic disease as part of a longer ongoing project. The researcher works diligently under the guidance of the project manager and group leader on an established grant, and presents a summary at the end of each week outlining their protocols.

After several months, other team members begin to apply the protocols in test-sample analysis, leading to a journal article submission. The junior researcher's name is omitted from the manuscript because they did not specifically run the test-sample assays. Understandably, the junior researcher is distressed that recognition for their work has been overlooked when it came time to publish, and that their contribution is not even acknowledged. They immediately begin looking for their next professional role with the intention of leaving early, taking the research concepts with them.

This case study represents one scenario where a team member feels discontent due to poor management and lack of acknowledgement. This type of scenario is more common than you might think, and it could have been avoided here with better communication, as outlined throughout this book.

Lessons Learned

In the private sector, the things you learn from each project experience are often recorded in a "**lessons learned log**," which can help to prevent you from repeating the same professional mistakes again. It can also prompt you to repeat something that has previously worked well in a later task or a future project. The same concept can absolutely apply to academic research projects, and is often done so informally. Indeed, cutting edge research operates with trial and error by nature. Nonetheless, it is rare to see researchers being trained to systematically follow a "lessons-learned" approach.

Lessons learned may reflect on positive or negative experiences that occur during the undertaking of a project. What is key is that whatever is learned from these experiences can be taken into account in future projects, or in later stages of the same project, and in a positive way. Positive adaptations may relate to improved research outcomes through superior experimental approaches, cost savings on project expenses, or ways to expedite the project's progress. All of these issues should be a focus of a project manager who constantly strives for excellence in the projects that they are overseeing. Similarly, the lessons learned may relate to issues that have been dealt with in the current project, and how to avoid them in future research pursuits. The project manager may also record lessons from certain approaches they've taken to managing the team, noting both things that worked well and other things that were not effective managerial approaches. This is particularly effective in larger project teams, and it provides another framework in which the project leader can demonstrate their work practices to the group leader. An example of a record that might be made regarding lessons learned is shown in table 14.2.

Table 14.2: Lessons learned

Name: Timothy Jones
Date: 5th October 2016
Activity or practice (issue): Set up cell proliferation assays as a team, instead of individually.
Impact: This new approach will improve the speed and accuracy of developing proliferation datasets.
Lesson/action: This week the project team decided to work together to set up their cell proliferation assays, as opposed to their usual practice of each running their own assays for their individually assigned test foci. The outcome was a very large-scale assay, which ran like clockwork and proved very efficient. By coordinating the team on this once weekly, they all achieved a solid dataset quickly and required less time altogether in the tissue culture facility, less waste of reagents (maximizing the cells available), and less cost overall. It is recommended to continue on with a weekly scheduled tissue culture assay set-up, coordinating together to split tasks.
Reference to lab book/other: P5-book2, page 57.

Each project participant can keep a personal log of lessons learned, or there may be a central location in which these lessons can be placed for all team members. The log may be kept in digital or paper format, or recorded on pages in an established notebook, if applicable.

Keeping a central record of lessons learned, which outlines improved practices that streamline or improve upon current practices, is particularly important for project teams that change over time due to various participants completing their contract and moving on to their next role elsewhere. Of course, if the lesson is specifically related to improving a protocol, the protocol itself should be directly modified and a new version created to share in the common drive or folder that the team keeps as a reference.

Recording Protocols

For experimentalists, **protocols are essential for establishing clear scientific practice**. In order to report on the outcomes of an investigation, the protocol must be clearly described in a manner that enables another researcher to reproduce the work and obtain the same results. Thus, recording clear protocols is critical in any experimental investigation, particularly when working on projects with the express goal of communicating the work in some form. For projects involving a team, the sharing of protocols ensures that all participants are aware of what is being delivered by other project participants, encourages feedback, and potentially aids the improvement of any approaches being taken.

It is good practice to keep a clear and open policy in your group regarding protocols in use. Do this to promote connectivity and support between all group members and project participants. **Establishing a central protocols database** makes the drafting of the final manuscript, grant, or patent a simpler process, because all protocols followed during the work should be able to be referenced back to the central protocol database. This complements referring to lab books and journals that record finer details, issues confronted, or reflections and observations on the day's experiments, where applicable.

CASE STUDY: Sharing protocols

A research team was operating within a larger established research group, involving several postdocs, two PhD students, and a technician. They were all working together on the same research foci, covering several principal projects underway at any one time. One of the postdocs was very competitive and sought to demonstrate that he was the best, and that he could achieve better outcomes than others in the group. The postdoc worked largely on his own, but did contribute to some project team datasets. The postdoc was not at all interested in contributing his protocols to the central database, because he viewed the detailed protocol approaches he had developed to be an extension of his skillset, and thus his own intellectual property, which was hard won and not to be shared lightly. The outcome was that the project did progress well, and a journal-style protocol description was made available by the postdoc. But once he left to commence his own group, the remaining group members were unable to repeat the work because they could not establish the key nuances behind making the protocols work effectively.

Over a longer timeframe — the life of a research group — demanding transparency and contribution to a central protocol database will strengthen the capacity of the group as people come and go. Newer group members embarking on projects in your group can benefit from the collective wisdom of all who came before them.

The above case study shows how poor collaborative practice in a research group can negatively impact project progress, despite all team members working under the same group leader. In this example, the postdoc refused to contribute to the collective development of skills for the team, which ultimately had two outcomes. The group was unable to replicate the work given the lack of clarification of technique or lessons learned. But the second outcome was that the postdoc would fail to create a positive professional network with the group members with whom he should be building a collaborative research project. This failure to participate in a team environment ultimately impacts the productivity of all involved, and wastes time and money that could otherwise aid the delivery of improved outcomes in a more timely fashion.

Communication and Improvement

Taken together, communication and a consistent focus on record keeping and on the enhancement of your team's approaches to projects are central to maximizing your project deliverables within the budget defined for the work. Many of the practices that benefit the project will be established in the group by the group leader, as we discussed in Section 1; by the group leader doing so, the effort required by the project manager will be greatly reduced. Establishing a group practice of recording lessons learned and research protocols in some kind of central database or shared drive can drastically improve productivity for all of your project teams, and overall will maximize on communication and collaboration for those working under you.

END OF CHAPTER 14 SUMMARY
Communicating and Improvement

In this chapter we discussed key issues relating to project communication, the personal communication goals of different project participants, and approaches that can aid consistent improvement of project approaches. The take-home messages for this chapter include the following:

- The overall goals and expectations regarding intellectual property and innovations stemming from the work should be made clear, and they should be openly communicated with all project participants at the outset.

- Updates and outcomes should be communicated efficiently to funding bodies and project participants.

- Presenting work at conferences and in journals benefits all of the project team, and this should be reflected in the communications stemming from the work.

- The right to adapt the work, or to utilize technology stemming from the work, should be openly negotiated between the group leader and the project participants.

- The project team should reflect on the work they have completed in order to maximize productivity and efficiency in future work activities by recording any lessons learned.

- Clear and open communication regarding protocols, and the establishment of a central protocol database or record, represents positive professional practice throughout the project's life.

REFLECT

1: Do you operate your projects with defined expectations regarding communication between all involved in the work?

2: Have you encountered challenges regarding technology development in your experience, and what was the root cause of the issue? How could you have addressed it better?

3: What official guidelines do you provide your team regarding the recording of their protocols and any improvements/issues they have addressed during the lifetime of their project? Did your guidance promote efficiency in your team?

EXERCISE 14.1: RECORDING PROTOCOLS

Petition your team to create an outline in an Excel file of all of the protocols that they utilize in their research. Establish a shared drive where each team member can record their protocols in a shared folder that everyone may access. Consider creating a special meeting every six months or so for the team to focus on updating and discussing their protocols, and consistently request that staff update their protocols in the shared drive.

EXERCISE 14.2: LESSONS LEARNED

Consider creating a "lessons learned" form that members in your team and group can use to record their advances in knowledge relating to their work. What is important that they record for your team? There are many resources available online to reflect on what types of headings or styles of forms you might use. Table 14.2 is downloadable from the Practical Academic website, both completed and blank, for you to use as a template that you may expand upon.

Lessons learned log

Name:
Date:
Activity or practice (issue):
Impact:
Lesson/action:
Reference to lab book/other:

Chapter 14 — Downloadable Materials
Download from www.practicalacademic.com

- Table 14.2 and exercise 14.2: "Lessons learned" — Excel file containing a blank template and the original table 14.2 as an example.

CHAPTER 15

Authorship and Publication Ethics

Although it is addressed throughout this book, the importance of authorship will be discussed here in its own right. Given that your contribution to work and overall productivity in academic research is typically gauged by your name appearing on research output, be it a patent, journal article, conference presentation, or thesis, **authorship is critically important**. Publication lists are typically used for performance review, academic promotion, and funding opportunities.

Disputes and ethical questions over authorship and ownership of ideas date back as far as recorded academic pursuits. Today, many academics continue to challenge contemporary academia's approach to authorship ethics and, in many cases, authorship practices may be related to a cultural or home-coussntry attitude toward group recognition. Similarly, discipline-specific publication practices can be as diverse as the disciplines themselves. The authorship debate is a persistent and ongoing concern in academic research.

Accountability, Responsibility, and Credit

Authorship on an academic paper conveys many things. It conveys not only that you contributed to the work, but also that you should be credited for the work that you have done. This is typically noted in the location of your name in the list of authors; however, there are varying practices to be found regarding how names are listed. These may depend on your field of interest, your home institution, or your target journal. Most critical, however, is that you are responsible for the content in a published article. If that article is proven to be carrying falsified data, if the procedures were performed unethically, or if it shows an inaccurate record of information, the authors are held responsible. You are responsible and accountable for all of the content to which you assign your name. Associate researchers have been known to request their name removed from an article submitted for publication because they felt it was inaccurate or unethical in some way, and they did not wish to be associated with said document. **Publication is a permanent record.**

> ### YOU ARE RESPONSIBLE AND ACCOUNTABLE FOR THE CONTENT TO WHICH YOU ASSIGN YOUR NAME

Types of Publication
Most publication-focused academic research will be targeted toward the development and publication of a peer reviewed journal article in a high-impact journal with a high citation index; however, there is a variety of concerns when identifying your target journal or publication output. These differences will be discussed here.

Discipline-Specific Authorship Styles
You should follow some basic guidelines when you define who should be recorded as an author on your paper. Nonetheless, gauging the relative contribution of authors remains an international source of confusion. Different disciplines and specific organizations will have their own approaches to defining authorship lists, and there are **no absolute clear guidelines**. Authors may be included to author lists under various circumstances discussed throughout this chapter. The organization of the authors' names on an author list may be via a

number of styles, which may or may not reflect their relative contributions to the work.

<u>List of authors styles</u>

Alphabetical: Authors are listed in alphabetical order. At some institutions, everyone at the lab at the time that the discovery was made is included in the author list, and in alphabetical order.

Strict author ranking: Biological sciences — unofficial
First author = main project leader/contributor.
Middle authors = may or may not have participated heavily in project (typically in order of relative contribution).
Last author = group leader — intellectual and financial driving force directing the research.

First and last author emphasis: Authors ranked by relative contribution according to the following overview.
First author = main contributor — 100%
Last author = second contributor — 50%
Other authors = overall impact divided by number of authors

Order of importance: authors ranked by relative contribution from first to last author.
First author = greatest contributor.
Second author = second-greatest contributor.
Third author = third-greatest contributor.
Last author = lowest level of contribution.

Equal credit to all contributors: indicates that all authors receive equal credit for the work. Overemphasizes the minor contributions of many authors.

Percent-contribution-indicated: the clarification of each author's actual contribution by percentage. (Tscharntke, 2007.)

Detailed/clarified: a footnote may be used to clarify the contribution of each author to the manuscript.

The method of authorship recognition may also be noted in the acknowledgements section.

Who Writes the Paper?

Whilst the first author is typically the dominant writer for a research article, and this is the accepted practice, other practices sometimes take place.

The first author

The first author typically coordinates and collects the data contributing to the paper, and is then predominantly the person who drafts the manuscript.

The research team

Each author contributes their results and, in some cases, the material related to their results. This is usually the practice for collaborative projects, where multiple contributions are coordinated by the first author, or by the group leader. Once the entire article is completed, all of the research team will revise the manuscript.

The group leader

The group leader drafts the paper at times, despite oftentimes not being the person performing the work that generates the data.

1) **The work may be done by a research technician** under the guidance of the group leader. In this case, the technician delivers their data back to the group leader for the latter to collate the work and draft the manuscript.

2) Another case in which the group leader drafts the paper may be when the **project leader demonstrates poor writing skills**, which is sometimes the case for international graduate students who speak English as a second language. This is particularly evident in groups that are under a tight schedule or wish to churn out papers quickly. Nonetheless, the practice of the group leader drafting the paper for a graduate student is ill-advised. Part of the graduate student experience is to draft their own papers to develop their skills as a researcher, and the paper should then be revised by the group leader, who should provide ongoing guidance.

3) The **group leader is a master orchestrator** in some cases. The group leader may also have an established operation in their group whereby the researchers actively pursuing the work are expected to pass their data/results to the group leader once a manuscript-worth of data is collected, at which point the group leader drafts the paper. In this case, the group **operates like a data-generating machine.**

Shadow authors / ghostwriters

Ghostwriting is generally questioned regarding ethical practice, whereby the authors take credit for a manuscript that they publish in their own names but for which they have not written the actual text for submission to a journal. In this case, writers are usually recruited as consultants to draft the manuscript based on a collection of information provided by the researcher or group leader. This practice is more often seen in the private sector, where the money spent is considered worth the time saved, and there it is considered to be a form of outsourcing. This is acceptable if the publication clearly states that the manuscript was drafted by an outside source; however, ghostwriting is widely frowned upon by academics.

Writing clear arguments and explanations of your research, and contextualizing your arguments to the current state of the art, is a central component of academic publishing and recognition. Thus the practice of ghostwriting continues to be controversial.

Document Revision and Editing

Document revision and editing for correct language is accepted for journal publishing, and it is often encouraged by reviewers in order to polish the manuscript to a point that it reads smoothly and has improved clarity. This should be restricted to language and line editing, and should have no impact on the academic content.

Document revision may be performed using Track Changes within an electronic manuscript, or performed manually by making markings on a printed manuscript. When junior researchers are developing their academic literacy, they should request manual revisions of their work. By entering the annotated changes themselves, they will become more aware of their typical grammatical issues, and thus will learn how to correct themselves more readily in the revision process. This can improve the junior researcher's writing skills overall, and so may reduce their future requirement for editorial assistance.

Self Citation

Many authors practice self-citation in order to boost their recognition. Self-citation is a good way to demonstrate the background to the current article, to clarify continuity. Some authors do excessively self-cite; it should be limited to relevant articles in order to illustrate the work being described.

Publishing and Platforms

Arxiv platform (https://arxiv.org/)

Arxiv is a platform run by the Cornell University Library for the publication of preprint articles in a range of disciplines. Its range includes mathematics, physics, astronomy, computer science, quantitative biology, statistics, and quantitative finance. Not all of the articles presented to the archive reach publication in a journal, and some are submitted to a journal in tandem. Although the article is not peer reviewed, moderators do gauge the relevance of the articles submitted. Further Arxiv platforms are being released to focus on other discipline foci, including Biorxiv (http://biorxiv.org/) for biology — run by Cold Spring Harbour — and SocArxiv (currently under development) for social sciences. Beyond Arxiv, you may also consider using Figshare (www.figshare.com).

Open access, or not?

Journals generally approach the publication of your work in one of two ways. They publish your material and retain the copyright over the work, charging readers to have access to the journal.

More recently, journals have adopted an increasingly widespread option for authors to pay for open access of their articles, suggesting that authors pay for the privilege of others reading their work. Some journals also operate in an entirely open-access style, providing free access to readers at no extra cost to the reader or author.

If you are seeking to understand the publishing practice of any specific journal, you might consider consulting the free service hosted by Nottingham University, SHERPA RoMEO. This can be accessed at http://www.sherpa.ac.uk/romeo/index.php. Here you can gain some ideas about the rules and regulations that any journal might have. The site states, "RoMEO is a searchable database of publisher's policies regarding the self- archiving of journal articles on the web and in Open Access repositories".

The peer review process

Given the cutting-edge nature of academic pursuit, the publication of new research advances, particularly in Science, Technology, Engineering, and Medicine (STEM), requires the review of peers in the field. These peers assess submitted articles to gauge the quality of the research findings, whether they are worth publishing as a new body of work, and to what extent the findings contribute to the field.

The peer review process is usually moderated by an editor. The editor selects the reviewers and gauges from the reviews whether the work is worth being included in the journal that they work for. In some circumstances, the inclusion of your work may be considered based on recommendation by somebody of value to the journal, such as the recommendation for publication pathway that PNAS (www.pnas.org) practices. Overall, this process should, in best practice, present a solid means by which only the best work is accepted. Peer review is not without its flaws, however it remains the standard approach for publication selection.

Predatory publishers

Predatory publishers have become a concern for academics in recent times, due to the increasing incidence of questionable, unethical, and often exploitative behaviors seen in some new publishing organizations. This is a significant issue, particularly given the assignment of the copyright of your materials once you agree to publish them. Predatory publishers may practice a range of problematic approaches, which are described here.

Spamming: unsolicited requests and invitations to use your material in a book, or to submit your research to a journal, which the publishers can arrange on your behalf.

Little or no peer review: Many predatory open-access publishers accept articles quickly, and with little quality control in the form of peer review. In this way, fake manuscripts can reach publication.

Poor clarity in the publication process: descriptions of fees for publishing an article in an open access journal may not be made clear until after the article is accepted and ready to be published, thus blindsiding the authors who then must pay unexpected fees.

Fakery: The journal may use a fake address or fake editors, or indicate that certain people are connected to the journal without their knowledge. Fake impact factors, or absence of impact factors, may be evident. Similarly, the journal may model itself on more established and highly regarded journals, thereby making their product appear to be above board and of high value.

Predatory publishers undermine the legitimacy of the practice of publishing in scientific journals. A list of noted predatory publishers is updated regularly by Jeffrey Beall — "Beall's List" — and it is a blacklist of publishers that should be avoided (https://scholarlyoa.com/publishers/). Conversely, a "white list" of above-board journals is operated by the Directory of Open Access Journals (https://doaj.org).

Predatory publishers also target **theses published by recent graduates**. They approach the graduate with email and phone inquiries to request that they publish their thesis as a digital book that they can market via online booksellers, paying a small percentage of any profit to the student that provided the material.

CASE STUDY: Predatory publishers

A new PhD student had just completed her one-year postgraduate research project, including a short thesis describing the work that she had performed. In the first few months of her subsequent PhD studies, she was approached by a publishing company. They inquired whether she would be interested in publishing her thesis as a book with them. Upon performing an internet search for the company name coupled with "predatory publisher," many results came back indicating the publishing company would actually publish the thesis online, and would skim all but a small percentage of the profit (which small percentage they would feed back to the author), and that they would claim the copyright over the content in the process.

When do You Include Someone in Acknowledgements?

Various acknowledgements should be made in your research publication. Acknowledgements may be made for a number of reasons, many of which are discussed here.

Acknowledgements

Those who revise the manuscript for you should be included to the acknowledgements section.

You may choose to include paid editors or revisers in this section, which indicates to the reviewers that it has undergone professional language revision.

Those who provided technical assistance or contributed somehow, but not enough to be a co-author may be acknowledged here.

Some researchers may help with the development of your work by providing a resource or assistance with a technique, but not sufficient enough to be considered as a co-author. The correct place to thank them for their assistance is in the acknowledgements section.

Those who influenced the research somehow

If you have a contributor who provided mostly ideas and suggestions to move the work forward, perhaps in a mentoring capacity or through purely collegiate discussions, they can be included in the acknowledgements, rather than in the list of authors.

"We acknowledge the valuable insights provided by *X. Person* regarding the overall research plan."

Equipment / resource providers

If you are reporting on a highly experimental body of research, you may have sourced reagents or resources that are not specifically acknowledged in the method section. In this case, you may wish to acknowledge someone who provided you with these reagents or resources in the acknowledgements section.

Funding provider

The funding provision for all authors to be able to contribute to the work should be included in the acknowledgements section of any manuscript deriving from the work.

"This research was supported by a [name of funding body] grant provided to [author initials]. [Postdoc initials] was funded by a [name of funding body] fellowship."

Unnecessary acknowledgements

Non-specific people

Non-specific people do not require acknowledgement. That is, if you are going to thank people for helping you with the work, they should be named, rather than thanking people "in general".

Reviewers

It is acceptable to acknowledge the anonymous reviewers of your manuscript, although they do fall into the category of "non-specific people" which is discussed above. They may be mentioned in the following way: "We wish to acknowledge and thank the anonymous reviewers of this manuscript for their constructive comments." This is not necessary, but if such a reviewer made a significant difference to the article, you may wish to acknowledge their anonymous contribution. Such contributions may include suggesting you add an experiment that you may have overlooked, or helping you to consider a point of view that you may not have considered previously.

Issues with Acknowledgements

People who do not wish to be acknowledged

You may acknowledge people that, for various reasons, wish to not be associated with the work, or generally wish not to be acknowledged. Thus it is important that you notify people that you intend to acknowledge them in your manuscript. Some consider it to be rude to not be notified about an acknowledgement made. This issue does not apply to sources of funding, which should always be acknowledged.

Issues with Authorship

Promiscuous authorship / honorary authorship

Pressure to have your name acknowledged in publications for professional recognition has promoted an increase in the practice of promiscuous authorship or honorary authorship (Strange 2008). In various circumstances, people may be awarded authorship despite not having contributed in any significant way to a paper. This is actually an insult to those who did contribute to the work, and does represent a form of fraud and scientific misconduct. Similarly, people are sometimes included onto papers to garner support or strengthen its regard by improving the credibility of the research.

Reasons for promiscuous authorships:

- legitimacy,
- honorary,
- mutual support,
- duplication authorship,
- ghost authors,
- general institutional practice, and
- group practice.

CASE STUDY: Institutional practice

A new junior group leader returned to their home country after postdoctoral training at a prestigious overseas institution. Once their group was established and the first journal article submissions were ready, it was made clear that the expectation was to include the name of the department head to the manuscript's authors, given that they were in the same discipline. The department head had not, however, played any part in the research itself, and they had not even read the manuscript.

CASE STUDY: Group practice

A junior postdoctoral fellow joined a new group, recruited from abroad on spec. The postdoc had an incredible amount of motivation to succeed in the new research pursuit after a very successful PhD. During their first postdoc year, they received advice only in lab meetings each week, and pursued their work entirely individually throughout the year. After completing their first project year, the postdoc drafted the first publication stemming from the work, coupled with a draft funding application outlining the projected future direction, in order to develop the work further and secure research funding. When delivering these drafts to the group leader, the postdoc was advised to include not only the group leader's name on the paper, but also the names of the senior postdoc and two long-term established technicians in the group. Further, the postdoc was expected to allow the group leader and senior fellow to submit the funding application in their names, so that the postdoc could be employed from the grant application the postdoc had conceived and written.

This case study on group practice outlines a classic case in academia, in which senior group members tend to focus on maximizing the productivity of all group members by supporting the career of established and longer-term group members by including their names on research papers derived from more junior group members. The logic is often given that the junior group member would not have made such progress with their work if the more senior group members had not provided the resources in place to be able to pursue and deliver their work. Nonetheless, this type of acknowledgement is essentially non-ethical acknowledgement of people that did not contribute directly to the work. Unless an author directly contributed to the practical and theoretical production of the paper or research plan, they should generally not be listed as an author, but rather included in the acknowledgements section of the paper and their lead-up work cited.

Not being recognized for your work

It is actually quite common for young researchers to be overlooked for their contribution to a project once they have already left the group. Disputes regarding ownership of a new project concept or direction are also often seen.

CASE STUDY: Authorship after leaving a group

A final year PhD student received a prestigious fellowship to complete postdoctoral training at a highly-regarded research institution. The student had dedicated every waking hour to training for several years to complete their PhD. The literature review of the thesis was an extensive overview of the field, and the student wanted to publish it as a review, but this idea was dismissed by the group leader. Despite moving on to a postdoc position, the PhD student's supervisor made the student wait a year for extra data before publishing. After the PhD student left, a new researcher took over the project to add an extra experiment to the draft manuscript. After the work was completed a year later, the new researcher tried to claim first authorship based on their completion of the final figure. The PhD graduate fought to retain first authorship on their PhD work, but ended up having to accept the new researcher as a joint first author. Later that year, the PhD supervisor and new researcher published a variation of the thesis introduction as a review, providing no credit whatsoever to the

---

This case study represents an all too common occurrence where a junior researcher moves due to contractual advancement, and consequently fails to receive acknowledgement by the people that remained in their host laboratory. Of course, some group leaders do make the effort to consider their staff and students even after they leave, but this is by no means standard. For this reason, those who want to maximize their output from a project — and ensure that their name is recognized — are advised to have everything submitted and finalized before moving on to their next posting.

From a group leader's perspective, failing to acknowledge their junior researcher's contribution results in poor long-term representation from those who should be able to vouch for you as a research leader. Further, the potential expansion of your collaborative network as a group leader will be vastly stunted should you fail to build a positive network with those that you have directly mentored through their postgraduate studies.

Being incorrectly recognized for your work

A researcher's publication history is representative of their professional productivity. Thus incidents that impact the recognition of contributions to publications, impact the career of the academic. **Omitting or incorrectly naming a contributor to the work**, as well as failing to have the researcher review and provide signed permission to publish under their name prior to submitting, is professional misconduct.

CASE STUDY: Incorrect recognition in published work

A postdoctoral researcher collaborated on a project with another group, contributing a significant body of data to strengthen an article that the collaborator had already drafted. The postdoctoral fellow was not notified when the paper was submitted, received no opportunity to review the manuscript prior to submission, and gave no permission for his name to be included on the article. When the postdoc finally received a proofed and reviewed copy from the journal, which was about to go to press, he finally realized that the collaborator had listed their name incorrectly, mixing up his middle and first names on the manuscript. Furthermore, the collaborative group had signed the postdoc's name digitally, as if he had seen the work and agreed to the early proofs. The postdoc was furious. He politely requested that his group leader and the collaborator correct this immediately. The postdoc also contacted the journal to request that his name be corrected. The journal printed a corrigendum to note the correct author name; however, the postdoc was required to pay half of a week's salary to pay for it. The group leader and collaborator refused to help.

The case described here occurred because the people responsible for submitting and publishing the work showed little consideration toward a co-author who made the work possible. In fact, the signing of a digital submission signature on the postdoc's behalf represents ethical misconduct. These types of issues are not uncommon, and they represent a lack of professionalism and respect. Given the competitive nature of the field and the ever-diminishing amount of funding, coupled with the abundance of qualified professionals vying for fewer positions, this type of lack of professionalism can impact a researcher's reputation. Further, it does little to help the group leader to build a network of professional contacts and likely collaborators, with productive and fresh new young group leaders emerging from training in their group. Instead, this type of behavior solidifies the opinion of the group leader who commits the misconduct as somebody that the postdoc in question would choose never to do "business" with again.

Plagiarism checking

Several plagiarism checkers are available online, and your institution may have access to certain platforms from which they pay for the service. These can prove useful, particularly if you are a group leader that promotes independent authorship practices by your first authors. In particular, given the extensive literature review required for many articles and theses, you might encourage your students to pass all of their written work through a plagiarism checker. This can avoid unnecessary embarrassments later on.

Theses are now often a matter of public record once published, and digital versions are often available online as publications in their own right, usually through the digital library of the institution from which a student graduated. Beware, however: **use an above-board plagiarism checker**. Consult your institutional research librarian for more advice on this matter.

Language checking

If you are not a confident writer, or if you are not a native speaker of the language you are publishing in, language revision is advised for you to ensure that your manuscript reads clearly and smoothly. A vast range of opportunities are available for you to have your manuscript revised through direct or online services. Such revisions are a wonderful way to help you improve your language skills in the process. Please ensure that you acknowledge that language revision has taken place.

Any manuscripts that have undergone language revision by a qualified language reviser should indicate it somewhere within the text. This is typically placed in the acknowledgements. (e.g.: "This manuscript has had the language professionally revised for publication." Or: "We gratefully acknowledge the language revision by Dr Jennifer Rowland.")

Authorship checklist

Here are a range of things that you might consider when you are deciding on the authorship list for an article.

- Have you provided data, a figure, or theory to the paper?
- Have you submitted any section of text, or revised the text for its academic merit?
- Have you had the manuscript revised for language fluency?
- Did you provide the conceptual framework and apply for funding for this work?
- Did you complete the bulk of the research work, contributing to more data/information than the other authors, and did you draft the main manuscript?
- Did you contribute to some of the data of the manuscript, and did you participate in discussions regarding the project? Did you perhaps contribute some of the text?
- Did you work together equally with another researcher to complete the work, sharing the responsibility of drafting the manuscript together?

Table 15.1 outlines a checklist that you might employ when considering how to decide on authorship for your next article. Clement (2014) recommended an authorship matrix approach that you may wish to take into consideration if you want to put some serious reflection and logic into your authorship decisions. This does not discount the traditions nor pressures that you are likely to encounter when it comes to authorship decisions, but it could certainly bring into focus the decisions as they are made in context, and it could help you and your group to create a fairer recognition of all parties involved in getting the research published.

Table 15.1: Authorship checklist

Project/Paper:_____

	\<Author initials>	\<Author initials>	\<Author initials>	\<Author initials>
Management				
Conceived project plan				
Drafted project plan				
Overall project mentoring				
Wrote funding application				
Contributed to experimental plan				
Managed project				
Research component				
Conducted experiments				
Analyzed data				
Prepared figures				
Completed statistical analysis				
Provided technical assistance				
Provided resources essential to work				
Provided collaborative data				
Presentation component				
Wrote/contributed text to first draft of paper				
Created argument/discussion				
Created literature review				
Wrote methods				
Created reference list and sourced materials				
Wrote abstract				
Contributed to conceptual interpretation of data				
OTHER				

List author's name and note their contribution to the project. A simple yes or no, or a percentage contribution to each task, may be considered.

END OF CHAPTER 15 SUMMARY
Authorship and Publication Ethics

In this chapter we have discussed issues relating to authorship and publishing academic research articles in peer-reviewed journals. This has focused predominantly on the accepted practices of authorship recognition, acknowledgement of contributors to a body of research, publishing platforms, and common issues that are confronted relating to these. The main take-home messages include the following:

- Transparency is paramount in authorship practices, particularly in relation to author lists on articles.

- Acknowledgements should be made to those who assisted the work, as well as to funding bodies.

- There are a range of things to take into consideration when you are publishing, including whether to go for open access or not, and you should take care when selecting a journal.

- Researchers should practice ethical publishing and authorship protocols.

- Consider the contributions of each individual author when deciding how you will structure your author lists for publication.

REFLECT

1: How do you decide who will be included in the author lists on the articles published by your group? Do you have an authorship policy or established guideline?

2: How do you support your team in their preparation of publications? Do you encourage your group to develop their own work? Does your team have a budget for document revision services?

3: Have you encountered any issues with authorship in the past? What was the problem? How did you resolve it? What impact did it have in the long run?

EXERCISE 15.1: AUTHORSHIP
Look at the last five papers that have been published by your group, and consider how you ranked the authorship for each of the contributors. Was it fairly equated? Was anyone omitted who should have been included? Was anyone included who should have been omitted? Complete table 15.1 to help you gauge the author contributions.

Downloadable Materials
Download from the www.practicalacademic.com

- Table 15.1 and exercise 15.1: "Authorship checklist" — Excel file containing a blank template.

Further Reading
There is a wealth of material in the public domain relating to authorship issues in academic circles, which is beyond the scope of this guidebook. Please see some specific resources listed here.

Clement, T.P. (2014). Authorship matrix: A rational approach to quantify individual contributions and responsibilities in multi-author scientific articles. *Sci Eng. Ethics*, 20:345-361. doi: 10.1007/s11948-013-9454-3

Dance, A. (2012). Authorship: Who's on first? *Nature* 489:591-593. doi:10.1038/nj7417-591a

Elliot, C. (June 5, 2012). On Predatory Publishers: a Q&A with Jeffrey Beall. Brainstorm. The chronicle of higher education. Online. http://chronicle.com/blogs/brainstorm/on-predatory-publishers-a-qa-with-jeffrey-beall/47667

Female Science Professor Blog. Web link. Acknowledgements: http://science-professor.blogspot.com.au/2009/03/authors-gratefully-acknowledge.html

Igou, E.R., and Tilburg, W.A.P. (2015). Ahead of other in the authorship order: names with middle initials appear earlier in author lists of academic article in psychology. *Frontiers in Psychology*, 6(469):1-9. doi: 10.3389/fpsyg.2015.00469

Kornhaber R.A., McLean, L.M., Baber, R.J. (2015). Ongoing ethical issues concerning authorship in biomedical journals: an integrative review. Int J Nanomedicine, 10:4837-4846. doi: https://dx.doi.org/10.2147/IJN.S87585

Mandal, M., Bagchi, D., and Basu, S.R. (2015). Scientific misconducts and authorship conflicts: Indian perspective. *Indian J Anaesth*, 59(7): 400-405. doi: 10.4103/0019-5049.160918

Marusic. A., Hren, D., Mansi, B., Lineberry, N., Bhattacharya, A., Garrity, M., Clark, J., Gesell, T., Glasser, S., Gonzalez, J., Hustad, C., Lannon, M-M., Mooney, L.A., and Pena, T. (2014). Five-step authorship framework to improve transparency in disclosing contributors to industry-sponsored clinical trial publications. BMC Medicine, 12(197)1-10. doi: 10.1186/s12916-014-0197-z

Noorden, R.V. (2013). Mathematicians aim to take publishers out of publishing. Nature News, 17 January 2013. doi:10.1038/nature.2013.12243

Strange K. (2008). Authorship: why not just toss a coin? American Journal of Physiology - Cell Physiology. 2008;295(3):C567-C575. doi:10.1152/ajpcell.00208.2008.

Smith,E., Hunt, M., and Master, Z. (2014). Authorship ethics in global health research partnerships between researchers from low or middle income countries and high income countries. *BMC Medical Ethics*, 14(42):1-8. doi: 10.1186/1472-6939-15-42

Tscharntke T, Hochberg ME, Rand TA, Resh VH, Krauss J. Author Sequence and Credit for Contributions in Multiauthored Publications. *PLoS Biology*. 2007; 5(1):e18. doi:10.1371/journal.pbio.0050018.

CHAPTER 16

Completing The Project

Projects are finite. They start and end, which provides an essential framework in which to operate effectively. Once the project is finished, there are a number of completion tasks to address in order to be able to draw a line under the current work and move on to the next project focus. As the group leader, it is important that you ensure that your project manager brings all of the required project issues to a close. In particular, if the project manager leaves and moves on to their next role, you need to ensure that future staff and students will be able to carry on the work with little difficulty. Thus, providing your project manager with clear expectations regarding completing the project is important in the promotion of the overall continuity of your group's research. The main matters to consider are discussed here.

What Defines Completion?

Several points can be used to clarify the end of a project.

- the project goals have been achieved,
- the project has run its course, or
- the project naturally rolls over into a subsequent project or a larger investigation.

These will be discussed throughout this section, particularly relating to how these scenarios impact the tasks that need to be completed at the end of a project.

Project Completion Goals

Ending a project facilitates a time to take stock of project achievements and experience, and provides the team with a chance to identify unachieved goals and gauge whether they should be studied further. The project manager should take charge of wrapping up the work, and of directing all of the tasks associated with finalizing the project as it closes. Such tasks include:

- tying up loose ends;
- identifying any new project directions stemming from the work (i.e. proposals);
- storing records and data;
- storing resources or products of the finished project; and
- ensuring all reports are delivered and updated.

Reporting

In various circumstances it is necessary to create reports at the end of the project, but it is also helpful in **defining the success of the work and gauging how effective the work was** in achieving the goals outlined in the original project plan (chapter 9). End-of-project reports are standard practice in academic project management, and they should be encouraged from all participants at the end of any project, regardless of whether the project is being carried forward to a new project/line of investigation or is completely coming to an end. Several types of end-of-project reports are discussed here.

Reporting to funding body

The final report that you deliver to your funding body will allow them to account for the expenditure of your work and to report to their investors or funding community regarding the outcomes of the investment. The structure and guidelines for final project reports are generally provided by the funding body, and are usually specific to its requirements. The final report will usually include a summary of the key deliverables and outcomes of the work related to the original project proposal. Explanations are provided regarding issues that were encountered and how they were addressed, in addition to how they impacted the overall outcomes.

Reporting to institutions hosting students

Numerous academic projects are tied to postgraduate students pursuing qualifications, in particular Doctorates or Research Masters/Honors. As such, institutions require summaries of the overall achievements made during the project period. This is usually delivered in tandem with the thesis to the university's higher degree research office (or equivalent), where it is kept on file. The provision of regular reports and the final report allows the institution to track the progress and outcomes of their students, and such reports usually also include any issues encountered and explanations regarding them. This may encompass multiple reports, including: student report, supervisor report, and a departmental report.

Final individual reports

Each project participant should complete their own individual report on their contributions to the work over the entire project, outlining the specific work that they contributed to, and noting the resources that they have developed and where they can be located/are stored (see below: "storage of project resources"). This may constitute a collation of intermittent reports that are made throughout the life of the project, along with an introduction and summary to frame the work. Several team members may report together if their work was closely related.

Overall report by project manager

A project manager report should be made to outline the overall project from start to finish. It should emphasize the deliverables that were met.

Similar to individual reports, the project manager's report should tie together all of the data generated, and may include other reports to define specific issues in more detail, like a dossier. The project manager's report should outline the work that remains to be addressed, and should allude to any future directions that may represent the next natural stage of investigation. All of the project completion tasks and reports should be recorded here.

Completion Tasks

Some major completion tasks that relate to experimental research projects in particular are discussed here, along with suggestions of how to best address them.

Storage of project resources

If a highly experimental project has come to a close, a number of resources that have been generated during the project will need to be stored. These resources should be stored well so that they are available for future work or potential collaboration stemming from the project's publications or other communications. Preparing a clear summary of the remaining project resources that are stored is recommended. These resources can be listed with some indication of where they can be found and where the details of that resource can be sourced, to ensure ease of retrieval at a later date.

One common gripe amongst researchers is the difficulty of finding resources from previous projects, which are either unable to be located, not located where they are supposed to be, or are left in such poor condition that they are unusable.

Ensuring that your project teams tidy up their resources and create a clear outline of their storage location and relevant details is critical for continuity in your research group. Table 16.1 provides an outline of the types of materials that might be stored.

Table 16.1: Storage of project resources

Material	Details
Animals	Species, genetic change (if applicable: type of stock (live, frozen embryo, etc.)).
Genetic clones	Novel constructs, commercial stocks, cloned DNA fragments.
PCR materials	Primers, buffers, reagents, controls.
Antibodies	Antibody stocks for various purposes, +/- fluorescent labels
Bacterial stocks	Bacterial strains for cloning, expression, and other purposes.
Cell lines	Cell culture lines for transfection, cloning, and carrying mutant constructs.
Synthetic chemical products	Type of synthetic chemical and purpose.
Samples	Biological/non-biological, field collection, other.
Prototypes	Materials developed for application.
Specialized equipment	Equipment specifically for experimental purpose.

CASE STUDY: Storage of project resources

A team completed a large investigation which involved the development of a number of novel transgenic mouse lines. The project was ending and not being carried forward for the foreseeable future; however, the transgenic lines may be adapted in further investigations over the next years. Before moving on, the group established a storage strategy for the mouse lines in which 2–3 founder males and females were kept for breeding as needed, requiring two cages. Before the mice reached two years of age, a mating would take place to ensure the living stock would not be lost. This approach ensured that new mutant stocks could be established relatively quickly. In addition, mutant sperm and embryos were cryopreserved in order to be able to revive the line for future investigations. Although this latter method would take longer, it served as a backup should the live stocks encounter difficulties in breeding.

The project team might use an outline with a framework similar to that shown in table 16.1 to form the cover of a dossier with further details of each individual resource. An outline is useful as an overall reference for scanning for useful resources at a later date, should the group move on to other work but need to utilize an earlier resource. These resources may be stored in central stocks established in the group, but keeping a record of those resources that were employed in a specific project provides a contextual framework of the previous application for the resource that will be applied. In particular, this gives future researchers an opportunity to look over how effectively the resource was used, and whether any issues arose in the work relating to it.

CLEAR CATALOGING OF RESOURCES WILL SERVE YOU WELL IN FUTURE PROJECTS

Even if the project is not ending at this stage and merely develops into a new project that logically follows on, it is wise to **store key project resources as a backup** in case the active resources fail for some reason. This is particularly critical with biological samples, such as cell lines, mutant mouse lines, or other stocks that are prone to contamination, or which are particularly sensitive and prone to degrading.

Recording backup resources

Individual **specification sheets** can complement the table outlining the resources stored from the project (table 16.1) to create a dossier of resources, to which you can refer after the project is completed. A mock example of a specification sheet for a plasmid is shown in table 16.2. Specification sheets should be collected throughout the project for reference, or may be adapted from experimental records in laboratory books. The sheet should include all of the key information required for the resource's use, where it can be located, what form it is stored in, where further information can be sourced, the date, and the name of the person creating the specification sheet.

Table 16.2: Specification sheet for plasmid

Plasmid name: pBS SK+ PGKneoNTRtkpA — selection cassette.
Plasmid description: Mouse phosphoglycerate kinase 1 promoter, neomycin resistance gene, non-translated region, thymidine kinase, polyA.
Reference: PNAS (1994), *91*(7), 2819-2823. JK Book 4, page 152.
Location: Freezer 4, shelf 2, box 132, position A4 **Stock preparation details:** midiprep, 10ug/ul

Collation of protocols

During the course of the work, your team will have developed and optimized a range of experimental protocols specific to the project underway, which may or may not have already been updated in the central database or lab folder. At the end of the project, you should ensure that each project participant finalizes a completely **updated**

collection of current protocols to create a collection specific to the project. These should be operational protocols, and not abbreviated in the style required for publication.

CASE STUDY: Group protocols

A research group focusing on immunology-based research hosts several postgraduates, postdocs, and technicians, who move through the group year after year. All perform similar experimental approaches, incorporating innovation as it arises, but generally in a supportive and collective fashion. At the end of each person's contract, the lab comes together to collate a recent book of protocols to present to the departing group member in a ceremonial farewell. This creates a sense of community and tradition in the group, and it promotes long-term connectivity as each person moves forward professionally.

This case study demonstrates how the collation of protocols need not represent a difficult task, but can contribute to the connectivity of a group and promote engagement beyond the contract of each individual group member or project participant.

Digital content

Depending on the type of work being pursued, a project may generate an immense amount of digital content that could be useful in the time following project completion. It may be useful not only for the project participants, but also for new people coming to work in the group in the future. Where and how this content is stored over the longer term should be resolved before the project participants move on from the current project. A range of likely digital content requiring attention may include:

- Digital images/videos — e.g. fluorescent microscopy images, video, satellite images, overnight observation video.

- Large datasets/screens — e.g. microarray data, data mining, and large database initiatives.
- Data analysis — e.g. FACS, PCR data, and behavioral data.
- Digital references representing key background information related to the project.

Facility shutdown

Parts of your research facility may be required to be shut down as the project draws to a close. This is particularly clear for those undertaking a novel line of investigation, or those that utilize novel facilities to expand on the datasets being delivered. If this is required for your project team, the project manager should ensure that this is scheduled so that the tasks required are addressed before the project is officially completed (see "project completion checklist" later in this chapter).

Final Lessons Learned

By the end of the project, you should have accumulated a number of lessons learned from the duration of the project (see chapters 12 and 14). Once the project is completed, the project manager should take some time to collate an overall **lessons learned summary**, reviewing all of the take-home advice that has been collated over the course of the project. This final review provides an opportunity to consider the best approaches that worked overall and those that failed, in order to approach the next body of work with a clear perspective of best practice for you and your team.

When the Project Never Ends

Projects often roll over into follow-on projects, or may form a part of a larger focal line of investigation. Thus many projects may be constructed around PhD-enrollment or grant-award timeframes, which may fit into an overall ongoing thematic investigation focus. More projects may continue directly on from the one that is defined to be now ending.

The current project may be coming to an end, but one or several spin-off investigation/s may already be planned or commenced. This usually has no impact on the reporting that needs to take place at the end of a project, but it can significantly impact the conclusion tasks, because materials may not need to be packed away for indefinite storage. Nonetheless, the end of a project and the rolling over to a new investigation should present a waymark where backups are made to ensure that significant resources are not lost as the work moves forward.

> **PROJECTS MAY ONLY REQUIRE A TURNOVER OF RESOURCES, BUT BACKING UP IS WELL-ADVISED**

Checklist of Project Completion

As the project is drawing to a close, it is important to set aside time to consider the final tasks specific to the work that need to be performed before commencing the shutdown process. An example of a project completion checklist is shown below, in table 16.3.

Table 16.3: Checklist for project completion

Final reports: - Project manager. - Funding body. - Institution. - Individual.

Final tasks:
- Storage of resources:
    - physical storage,
    - overview/outline, and
    - specification sheets.
- Protocol collation.
- Digital content collection.
- Facility shutdown.
- Final lessons learned.

Completed Project

Once all of the project completion reports and tasks have been written and performed (while being overseen by the project manager), the reports should be delivered to the group leader to keep as reference material moving forward. Depending on the confidentiality of the work and the agreements made regarding ownership and permission to continue the work moving forward, the project manager may keep a duplicate of this report, and may also duplicate the samples and digital content from the shutdown process.

END OF CHAPTER 16 SUMMARY
Project Completion

In this chapter we discussed the processes that are generally followed at the completion of a research project, and under the guidance of the group leader and project manager. Specific completion tasks and reports that should usually be completed are discussed, and an example of items that might be considered is provided. The main matters described were as follows.

- Ensure all project reports are delivered from participating researchers and the project manager, and to the appropriate parties (funding body, institution, and so on).

- Ensure all resources are correctly stored and recorded for access in later work.

- Collate all protocols specifically related to the work as a key reference for later work.

- Ensure all digital content is appropriately backed up to secure locations.

- Shutdown the facilities that will no longer be required for foreseeable further investigations in the group.

- Create a summary of the lessons learned throughout the project, along with take-home messages.

- Complete a checklist of all of the completion tasks that need to be addressed specific to the project.

REFLECT
1: Do you keep clear guidelines for your group members to complete their work and hand it over?

2: Do you spend a lot of time chasing up information from projects that are long finished, in order to be able to utilize materials or information stemming from that work?

3: Do you try to maximize on professional efficiency by considering lessons you've learned over the course of the work you've performed, and do you share that knowledge amongst your team?

EXERCISE 16.1: PROJECT RESOURCES SUMMARY
Make a summary of all of the project resources in your group that are worth tracking by the research teams pursuing investigations under your supervision. Consider the materials outlined in table 16.1 and make your own summary of what needs to be tracked in your own research investigations. A blank spreadsheet can be downloaded from www.practicalacademic.com.

Material	Details

EXERCISE 16.2: SPECIFICATION SHEETS

Create an outline for your students and staff to keep track of resources that they are employing or creating during their current investigations. What materials do you think should be included in a clear inventory? An outline for specification sheets is available to download from www.practicalacademic.com.

Resource name:
Resource description:
Reference:
Location: **Preparation details:**
Further Details

EXERCISE 16.3: PROJECT COMPLETION CHECKLIST

Consider the projects that you currently have underway in your group. For at least one of them, create a project completion checklist that you think may encompass all of the considerations the project manager will be required to address before moving on to the next body of work.

Task	Person Responsible	Due Date	Notes
Final Reports			
1			
2			
3			
4			
Materials Storage			
1			
2			
3			
4			
Protocol collation			
Digital content			
Facility shutdown			
Final Lessons Learned			

Chapter 16 — Downloadable Materials
Download from the www.practicalacademic.com

- Table 16.1 and exercise 16.1: "Project resources spreadsheet" — Excel file containing a blank template, a completed example exercise, and table 16.1, "storage of project resources".
- Table 16.2 and exercise 16.2: "Specification sheets" — Word file containing a template for a specification sheet, and table 16.2 as an example.
- Exercise 16.3: "Completion checklist" — Excel file template for the creation of a project completion checklist.

REFERENCES – Chapter 16

Great Britain. Office of Government Commerce. (2005) Managing successful projects with Prince2.

Wu, H., Liu, X., & Jaenisch, R. (1994). Double Replacement: Strategy for Efficient Introduction of Subtle Mutations into the Murine Col1a-1 Gene by Homologous Recombination in Embryonic Stem Cells. *Proceedings of the National Academy of Sciences of the United States of America, 91*(7), 2819-2823. Retrieved from http://www.jstor.org/stable/2364327

Rowland JE et al., 2015 Mol. Cell. Biol. vol. 25(1), 66-77 doi: 10.1128/MCB.25.1.66-77.2005

SECTION 2 SUMMARY
Eight Things You Can Do To Improve Your Direction Of Research Projects

Here we have discussed a number of topics related to managing your research projects effectively. These are the take-home messages from each of the eight chapters of Section 2.

1: Be SMART in designing your project. Develop a clear proposal that addresses all of the key issues, a solid budget, and comprehensive ethics and safety considerations (chapter 9).

2: Include the right project participants in the work, ensuring that your expectations align with the requirements of the person you may include in the team. (chapter 10).

3: When starting the project, ensure the project manager establishes the work effectively, following key project startup requirements (chapter 11).

4: Once projects are underway, ensure that the project teams meet regularly, keep solid records, and provide reports when required (chapter 12).

5: Proactively pre-empt stumbling blocks in the work and anticipate potential change in order to best navigate advances and setbacks that arise during the course of the project (chapter 13).

6: Ensure that regular and open communication is maintained by all of the project participants in order to best advance the work, and to minimize any issues that may arise (chapter 14).

7: When publishing the work, follow ethical guidelines for authorship and acknowledgement of project contributors (chapter 15).

8: When the project comes to an end, ensure that all of the project materials are appropriately stored, reports are completed, and information is cataloged appropriately (chapter 16).

SUMMARY

Thank you for reading through this entire book! I hope that you have found some useful information in your exploration of the concepts proposed. In **section one**, these concepts have covered general approaches that you might take in your day-to-day management of your research group and overall responsibilities in your role as a tenured academic. **Section two** focused on how you might lead a project from conception to conclusion, suggesting management strategies you could use to ensure that you maximize on your project-derived productivity moving forward.

If you have worked through the **collection of exercises** presented at the end of each chapter, you should have now developed a very clear perspective of exactly how you perceive your current research "business" and have taken some time to reflect on your group, their strengths, and ways in which you can support ongoing and systematic advancement of your people and your research overall.

This book represents the first in what is intended to be the Practical Academic series of books designed to support research academics internationally. I encourage you to provide feedback in the contact form on the website (www.practicalacademic.com) or via other non-biased online feedback platforms. If this book has helped you in any way, I'd like to know how, and what aspects of the material most supported you. Similarly, if you felt it could be improved upon, please do let me know, so that I can improve the material in later titles and editions.

With thanks and warm regards,

Jennifer Rowland

APPENDIX 1: Some Virtual Resources Currently Available for Networking (2016)

<u>Virtual representation of your productivity</u>

Institutional and Research Group Webpage

Most institutions provide a website that you may utilize to outline the basics of your research focus and the members of your team. Expanding this to a site specifically designed to suit your research focus is an advisable way to digitally represent your work and publicize your group. You may include:

- links to your publications and media releases;
- your group's advancements;
- a bio of each group member, coupled with their specific research focus and their CV;
- notifications of new grants and awards received;
- links to relevant institutional pages;
- opportunities for employment or student positions;
- seminars, conferences, and courses relevant to your work;
- your current research activities;
- photos of the people and work in daily lab life; and
- an overview of the lab schedules for group members and potential candidates to refer to.

It is strongly advised that you incorporate the cost of developing a website for your group into your running budget. It will reap dividends for you.

LinkedIn (www.linkedin.com)

LinkedIn.com currently provides a free-of-charge digital professional networking tool, which is accessible to anyone and allows them to view your overall professional credentials once you have completed your profile. Proactive professional academics maintain an active and updated LinkedIn profile, not only to be able to market themselves professionally, but also to attract potential employees and students by demonstrating their excellence in the professional space. On your LinkedIn profile, you might include the following information about yourself:

- professional roles,
- unpaid experience,
- professional qualifications,
- publications,
- awards,
- endorsements from your network, and
- interests.

You can also use LinkedIn for searching for new jobs, or to try to be recruited for a new role. You can extend your LinkedIn profile to include a great deal of information, or you can be selective to maintain your privacy.

ResearchGate (www.researchgate.net)

ResearchGate is a free-of-charge digital professional networking tool, predominantly used by scientific researchers. This network presents similar characteristics to LinkedIn and other social networking sites where users make a profile. However, all ResearchGate users must be affiliated with a recognized institution or be confirmed via the moderators. Some benefits include:

- networking your work and publications,
- networking with other researchers,
- putting questions and queries to other researchers,
- showing statistics on your uploaded publications,
- recruiting and job searching,
- advertising for potential students and staff, and
- using it as a blog.

This forum is cross-discipline, but it is heavily used by biologists and medical researchers.

Academia.edu

Academia.edu is a free-of-charge digital professional networking tool, similar to ResearchGate, offering a very similar networking experience to that noted above for ResearchGate.

Fellowship Networks

Various fellowship networks will be available to academic researchers. They are typically associated with being a member of an organization, or with funding by a fellowship, making the fellow eligible to participate in an associated network. These networks provide opportunities to develop a professional network relative to your disciplinary focus.

Discipline-Specific Networks (DSNs)

Discipline-specific networks represent a collection of researchers that focus on the same disciplinary field. Larger DSN groups often come together in regular — typically annual or biannual — meetings and conferences. Smaller groups may be only a collection of several research groups working together on a focal research topic. In any case, DSNs often connect via virtual groups, which can be dynamic in structure. These may include an email list, blog, or online forum.

Working Groups

Similar to DSNs (above), working groups represent a collection of researchers that focus on the same overall project goals. Working groups typically focus on projects, usually a collection of collaborative projects, between researchers focused on achieving project deliverables. Participants can often be located in geographically diverse locations, and thus require effective virtual networking to communicate. Working groups usually employ email, online data sharing (such as Dropbox, Google Drive, or institutional file sharing),

Skype, or Zoom videoconferencing. They may also use phone apps, such as Asana, Slack, and Evernote. Asana tracks the project's progress and facilitates team conversations. Slack is a messaging app for teams that promotes fast messaging and file sharing, and it is similar to other popular messaging apps like WhatsApp and Viber. Evernote and Google Drive work together to provide project group communication, planning, and file sharing.

Online Courses
Coursera, https://www.coursera.org/
edX, https://www.edx.org/
Lynda.com, https://www.lynda.com/
Udacity, https://www.udacity.com/

Activity: make a list of the virtual networks that you participate in.

APPENDIX 2: Group Leader's Toolkit

The tables and exercises presented throughout this book provide a collection of resources that may be collated into a useful "toolkit" that group leaders may utilize as they manage their research. The toolkit is compiled here for ease of reference, and all source files are accessible via the Practical Academic website (www.practicalacademic.com).

"Map out your own research business" - table and exercise 1.1
"Outline your institutional services" - table and exercise 1.2
"Defining your research goals" - table and exercise 1.3
"Mind map your departmental structure" – no download

Your research team - exercise 2.2
Management approaches - exercise 2.1

One-page mentoring checklist - table 3.1
Employment history record- exercise 3.1
Personal and professional development (individual strengths and goals) - exercise 3.2

Ordering spreadsheet - table 4.1
Duties roster for research group - table 4.2
Funding sources - exercise 4.1
Databases - exercise 4.2

Networking within institution - exercise 5.1
Providing incentive - exercise 5.2
Institutional support - exercise 5.3

Meeting schedule outline - exercise 6.1
Reporting - exercise 6.2
Planning for disruption/delay - exercise 6.3

Personal goals and characteristics questionnaire – table 7.1
Code of conduct – table 7.2 and exercise 7.1
Ownership of the work – exercise 7.2
Conflict management history – exercise 7.3

Value added – table 8.1 and exercise 8.1
Teaching - exercise 8.2

Project Planning - exercise 9.1
GANNT chart for project - exercise 9.2

Staffing table - exercise 10.1

Start-up checklist - table 11.1 and exercise 11.1

Journal club log - table 12.2
Communication - exercise 12.1
Meetings - exercise 12.2
Reports - exercise 12.3

Checklist for project change - table 13.2 and exercise 13.1
Troubleshooting and change - exercise 13.2
Project networking - exercise 13.3

Communication guidelines - exercise 14.1
Lessons learned - table 14.2 and exercise 14.2

Authorship checklist - table 15.1 and exercise 15.1

Project resources spreadsheet - table 16.1 and exercise 16.1
Specification sheet templates - table 16.2 and exercise 16.2
Completion checklist - exercise 16.3

INDEX

www.ingramcontent.com/pod-product-compliance
Lightning Source LLC
Chambersburg PA
CBHW060813220326
41598CB00022B/2604